OECD Studies on Environmental Policy and Household Behaviour

Greening Household Behaviour

OVERVIEW FROM THE 2011 SURVEY
REVISED EDITION

OECD
BETTER POLICIES FOR BETTER LIVES

This work is published under the responsibility of the Secretary-General of the OECD. The opinions expressed and arguments employed herein do not necessarily reflect the official views of OECD member countries.

This document and any map included herein are without prejudice to the status of or sovereignty over any territory, to the delimitation of international frontiers and boundaries and to the name of any territory, city or area.

Please cite this publication as:
OECD (2014), *Greening Household Behaviour: Overview from the 2011 Survey – Revised edition*, OECD Studies on Environmental Policy and Household Behaviour, OECD Publishing.
http://dx.doi.org/10.1787/9789264214651-en

ISBN 978-92-64-21464-4 (print)
ISBN 978-92-64-21465-1 (PDF)

Series: OECD Studies on Environmental Policy and Household Behaviour
ISSN 2308-1376 (print)
ISSN 2308-1384 (online)

Revised edition
Details of revisions available at:
www.oecd.org/about/publishing/Errata_Greening%20Household%20Behaviour-ENV-V2-ys%20(3).pdf.

The statistical data for Israel are supplied by and under the responsibility of the relevant Israeli authorities. The use of such data by the OECD is without prejudice to the status of the Golan Heights, East Jerusalem and Israeli settlements in the West Bank under the terms of international law.

Corrigenda to OECD publications may be found on line at: *www.oecd.org/about/publishing/corrigenda.htm.*

Preface

Governments everywhere are taking steps to encourage people to take environmental impacts into account in their daily life and purchases. But do these measures have any measureable effect? Are they leading to more sustainable consumption patterns? And how does environmental behaviour differ across households?

Developing strategies that promote greener lifestyles requires a good understanding of what things affect people's behaviour. To provide policy makers with some guidance on how to design environmental policies that are effective in influencing households, a project on Household Behaviour and Environmental Policy was initiated by the OECD in 2005. The work is based on large-scale periodic household surveys designed to shed light on household environmental behaviour and pinpoint how policies implemented by governments may affect household decisions in order to guide policy. The surveys focus on five areas in which household consumption has significant environmental impacts: energy, food, transport, waste, and water. Each subsequent round of the survey also allows behavioural changes to be tracked over time and to explore new and emerging issues.

The latest survey was conducted in 2011 and covered more than 12 000 households in 11 countries: Australia, Canada, Chile, France, Israel, Japan, Korea, the Netherlands, Spain, Sweden and Switzerland. The findings clearly underline the significant role environmental attitudes and norms play in shaping behaviours. Providing the right financial incentives is a key to influencing environment-friendly choices. In areas such as energy, water and transport, the provision of services and infrastructure can be an important complement. For specific groups of households that face barriers to the uptake of more environment-friendly practices, the provision of grants may also be needed. Stimulating desirable behavioural changes ultimately requires a mix of policy instruments.

Making the environment a priority starts at home. At a time when in many countries governments are adopting policies to reduce their deficits, stimulate the economy and create jobs, it is promising to see that approximately 70% of the respondents to this OECD survey agree with the statement that "protecting the environment is a means of stimulating economic growth". Through intelligent policy design environmental and economic objectives can go hand-in-hand. The OECD is helping countries to identify the policies which can make this happen.

The findings of the latest survey offer fresh insights into the sorts of policy interventions that are likely to work best. They build on experience gained from comparison across countries, environmental domains and over time. This book should appeal to all those interested in the challenging question of how we can promote greener consumer behaviour.

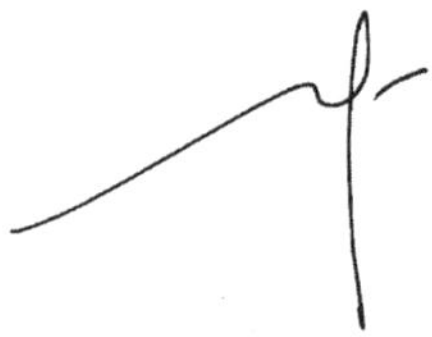

Simon Upton

OECD, Director of Environment

Foreword

Household consumption patterns and behaviour have a profound effect on stocks of natural resources and the quality of the environment. The importance of taking the "demand side" into account is a key lesson arising from the OECD's Green Growth Strategy. Governments of OECD member countries have introduced a wide variety of measures to encourage people to take environmental impacts into account in their purchases and practices. These may include environment-related taxes, energy efficiency standards for homes and appliances, fuel economy standards for vehicles, CO_2 emission labels for cars, and financial support to invest in solar panels. Nevertheless, influencing households remains a challenge for policy makers. Developing growth strategies that promote greener lifestyles requires an improved understanding of the consequences of such policy measures on households' decisions.

In an effort to develop evidence-based policy guidance, the OECD has implemented a periodic survey of households. This represents a breakthrough by providing comparable data on household environmental behaviour across a number of OECD countries. Based on responses from more than 12 000 households across a number of countries and five thematic areas (energy, food, transport, waste and water), analysis of the survey data offers new insights into what factors affect people's behaviour towards the environment and on what policy measures really work to enable change at the household level.

The second round of the OECD Survey on Environmental Policy and Individual Behaviour Change (EPIC) was implemented in early 2011.[1] *This report provides an overview of the survey data and some results arising from the analysis of the survey responses.*

The 2011 questionnaire was developed with the inputs from the Advisory Committee composed of national experts. It is largely based on that used in the 2008 round to ensure some comparability. However, refinements were made and new areas explored: eco-innovation, knowledge, policy preferences and country-specific questions. As in the 2008 survey, information was collected on household characteristics (e.g. age, income, education), environmental attitudes (e.g. concerns for the environment) and policy factors influencing household behaviours in the five areas considered such as the use of economic incentives (e.g. waste charges, grants to buy alternative-fuel cars) or eco-labels (e.g. energy efficiency of appliances and buildings, organic food). The full 2011 EPIC Survey questionnaire is provided in Annex A in English (Canadian). The links to the electronic versions of the online questionnaire implemented in the eleven countries surveyed are available in Annex B.

The two rounds of the EPIC Survey were implemented using the Internet and responses to the questionnaire were collected by means of online household panels in different countries. For representativity, the sample was stratified in each country according to different parameters: age, gender, region and socio-economic groups.[2] The 2011 survey results are based on a sample of more than 12 000 respondents in eleven countries, compared to over 11 000 respondents in ten countries in 2008. Six countries were involved in the two rounds (Australia, Canada, France, Korea, the Netherlands and Sweden), and five new countries took part in the 2011 survey: Chile, Israel, Japan, Spain and Switzerland.[3] More details on the survey implementation are provided in Annex B, including the selection of the survey provider, the technology used, the quota sampling and the response times.

As in all surveys, there is potential for sampling bias, in spite of the rigorous efforts made at stratification and quota sampling. The degree to which the country-level samples are representative of the national population is presented for a number of key variables in Annex B, and practitioners wishing to use the statistics and data herein are invited to review that material.

In addition, readers should note that this survey elicited respondents' stated preferences and perceptions. Therefore, statistics reported here which relate to objective, verifiable indicators (e.g. whether or not policy X exists in country Y) should be interpreted bearing in mind lack of public awareness about these indicators. For example it is possible that some respondents may not be aware that a given policy exists in their country. Similarly, some respondents may mistakenly believe that a policy exists in their country, when in fact it does not. However, it is important to note that for all questions in which there was significant potential for such "measurement error", respondents were given the option to respond that they "did not know" if such a policy was in place.

In general, readers should view these data as exactly what they are: the self-reported behaviours, attitudes and perceptions of representative samples of households from eleven OECD countries. Bearing the limits of such data in mind, it is important to recognise their advantages: information on households' knowledge and perceptions about environmental issues – increasingly recognised as a crucial factor for better understanding behavioural responses to environmental policies – is rarely analysed at such level of detail as in the following chapters. Moreover, for many variables such as discrete choices about whether or not a given purchase has been made, there is likely to be very little deviation from a more formal household consumer survey.

This book presents a first picture of the 2011 OECD EPIC Survey data and initial findings using descriptive statistics. It also sheds useful lights on key issues to be further examined as part of the follow-up phase of the activity on household behaviour where the dataset analysis will be refined by using econometric techniques in an attempt to better understand the drivers of environmental behaviour and the determinants of change towards greener behaviour. This analysis will complement the preliminary findings. The result of this new work will be synthesised in a forthcoming publication. The implementation of the third round of the OECD EPIC Survey is scheduled for early 2014.

Notes

1. The first survey was carried out in 2008 in ten countries and the main results were presented in the OECD publication *Greening Household Behaviour: The Role of Public Policy* (2011).
2. The quota targets relative to the samples, by country, are provided in Annex B.
3. It should be noted that the same respondents cannot be targeted over the years.

Table of contents

Acknowledgements 21

List of acronyms 22

Executive summary 23

Chapter 1. **The environmental policy context**

by Ysé Serret-Itzicsohn, Zachary Brown *and* Nick Johnstone 27

1. Environmental policies targeting household behaviour 28
2. The use of charges 31
3. The use of grants to encourage households to invest in eco-friendly equipment or products 33
4. The use of eco-labels 40
5. Availability of environment-related services 45
6. Household attitudes towards environmental policies 49

Notes 52

Chapter 2. **General household attitudes towards the environment**

by Zachary Brown, Nick Johnstone *and* Ysé Serret-Itzicsohn 55

1. Using general attitudes and beliefs to design environmental policy 56
2. Perceived importance of environmental concerns relative to other global issues 58
3. Perceived seriousness of specific environmental concerns 60
4. General trends in environmental attitudes 63
5. Clusters of environmental attitudes across countries and correlation with household demographics 66
6. Respondents' satisfaction/dissatisfaction with aspects of their local environment 70
7. Knowledge and beliefs about climate change 71
8. Conclusions 74

References 75

Chapter 3. **Household behaviour and energy use**

by Bengt Kriström 77
1. Introduction 78
2. Households' electricity consumption and spending patterns 80
3. Households' energy choices in their residence 85
4. Energy efficiency investments and behaviour 92
5. Willingness-to-pay to use renewable energy 102
6. Conclusions 107

Notes 108
References 110

Chapter 4. **Household behaviour and transport choices**

by Claude Weis *and* Kay W. Axhausen 113
1. Introduction 114
2. Overview 115
3. Clustering households by their environmental concerns 117
4. Car ownership 119
5. Car use 130
6. Household choice of transport mode for frequent trips 135
7. Support for government policies to reduce vehicle CO_2 emissions 143
8. Conclusions 144

References 146

Chapter 5. **Household behaviour and water use**

by Quentin Grafton 149
1. Introduction 150
2. Research on the drivers of water conservation 152
3. Overview 153
4. Analysis and results 157
5. Preliminary policy implications 173
6. Conclusions 176

Notes 177
References 177

Appendix 5.A1. Definition of variables 179

Chapter 6. **Household behaviour and food consumption**

by Katrin Millock *and* Céline Nauges 183
1. Introduction 184
2. Organic food consumption 185
3. Factors that would encourage consumption of organic food 193
4. Food waste, food "miles" and animal welfare 196
5. Environmental labelling and trust 199

6. Willingness-to-pay (WTP) for organic produce and animal welfare 204
7. Conclusions 211

Notes 214
References 215

Chapter 7. **Household waste generation, recycling and prevention**
by Ofira Ayalon, Sharon Brody *and* Mordechai Shechter 219
1. Introduction 220
2. Research on the impacts of waste policies and the role of households' characteristics 222
3. Waste generation 223
4. Disposal of waste containing hazardous materials 229
5. Waste separation and recycling 231
6. Attitudes towards waste management policies 238
7. Conclusions 240

Notes 243
References 243

Chapter 8. **Household attitudes across environmental domains and time**
by Nick Johnstone, Zachary Brown *and* Ysé Serret-Itzicsohn 247
1. Willingness-to-pay for different "environmental" goods 248
2. Reported motivations to conserve resources (energy and water) 251
3. Recognition of labels 253
4. Stated and actual behaviour 254
5. Households' adoption of technological innovations 256
6. Comparison of selected responses from the 2008 and 2011 surveys 257
7. Conclusions 258

Annex A. OECD 2011 Survey: Questionnaire 261
Annex B. OECD 2011 Survey: Implementation 291

Tables

1.1. Examples of measures targeting household behaviour 30
1.2. Use of eco-labels in the countries surveyed 40
2.1. Percentage of respondents in agreement with seven attitudinal statements, by class membership 67
3.1. Estimated average electricity price, by country 84
3.2. Reported annual expenditures on electricity, by recognition of energy efficiency labels 100
4.1. Main effects of socio-economic and attitudinal variables on car ownership and use 145

5.1. Water charge types (as percentage of total responses) 154
5.2. Correlation coefficients between water-saving behaviours and selected variables 161
5.3. Correlation between investing in water-saving devices and selected variables 165
5.4. Correlation between investing in water-saving devices and labelling 167
5.A1.1. Correlation between receiving water conservation grants and selected variables 180
5.A1.2. Correlation between satisfaction with quality of tap water and selected variables 181
5.A1.3. Correlation between drinking tap water and selected variables .. 181
6.1. Factors that would encourage consumption of organic food (2011 survey) 194
6.2. Factors that would encourage consumption of organic food (2008 survey) 195
6.3. Usefulness of information on animal welfare 195
6.4. Opinions on relationship between food consumption and environment 196
6.5. Median percentage of food that is thrown away 197
6.6. Food waste and organic food consumption 198
6.7. Organic food and animal welfare labels 201
6.8. Mean percentage expenditure and European label recognition and trust 203
6.9. The Marine Stewardship Council label (MSC) for seafood from sustainable fisheries 204
6.10. Mean willingness-to-pay and European label recognition and trust 209
6.11. Correlation coefficients between expenditure and willingness-to-pay 210
6.12. Reasons for not wanting to pay extra 210
B.1. Project timeline 291
B.2. Number of respondents per country 292
B.3. Screened, drop-outs and completions, by country 295
B.4. Questionnaire completion time and speeders 296
B.5. Quota targets relative to samples 298

Figures

1.1. Percentage of respondents reporting being charged on a unit basis for different environmental services 31
1.2. Percentage of respondents reporting having invested in eco-friendly equipment who received grants 33

1.3. Car buyers who reported benefiting from a scrapping scheme and/or financial incentive to buy a "green" car over the past five years 36
1.4. Recognition of different eco-labels 41
1.5. Reported access to selected environment-related infrastructures 45
1.6. Reported provision of environment-related electricity service options 47
1.7. Support for different policies to address vehicle CO_2 emissions .. 49
1.8. Support for different policies to reduce household waste generation 49
1.9. Support for unit-based waste charge policies by exposure to such policies 51
2.1. Respondents' three most prioritised world issues, by country.... 58
2.2. Percentage of respondents ranking environmental issues among the three most serious 59
2.3. Top three environmental concerns indicated to be the most serious 61
2.4. Relationship between general environmental concerns and specific concern for climate change 62
2.5. Respondents reporting "climate change" as the most serious environmental issue facing the world, by age and country 62
2.6. Levels of agreement with seven statements about environmental policy 63
2.7. Scepticism about environmental impacts and views on the seriousness of climate change 64
2.8. Views on intergenerational equity across ages 65
2.9. Agreement with the statement "I am not willing to do anything about the environment unless others do the same", by participation in volunteer organisations 66
2.10. Size of environmental attitude clusters, by country 68
2.11. Size of environmental attitude clusters, by respondent characteristics 69
2.12. Levels of dissatisfaction with local environmental quality....... 70
2.13. Dissatisfaction with air quality and household location......... 71
2.14. Respondents who believe that human activities contribute to climate change, by level of post-secondary education 72
2.15. Respondents who believe that human activities contribute to climate change, by level of trust in scientific experts 73
3.1. Distribution of households' reported electricity spending, by country 81

3.2. Reported annual electricity expenditures, by household size 82
3.3. Relationship between household electricity budget share and income, by country 83
3.4. Relationship between average electricity price and electricity consumption 85
3.5. Reported energy sources for space heating and cooling, by country 86
3.6. Reported installation of solar panels 87
3.7. Reported provision of special electricity services 88
3.8. Reported access to smart metering and interest expressed, by country 88
3.9. Reported demand for differentiated electricity rates, by country 89
3.10. Percentage of households reporting that they would make a "free" switch of electricity provider to benefit from "green tariffs", by country 91
3.11. Energy-saving behaviour index by country 93
3.12. Energy-saving behaviour by practice 94
3.13. Percentage of respondents reporting taking into account energy costs when changing residence 95
3.14. Reported energy-saving investment by home ownership status .. 96
3.15. Reported receipt of energy efficiency grants, by income category 97
3.16. Recognition of energy efficiency label: appliances and buildings .. 98
3.17. Respondents reporting taking into account energy costs when changing home 99
3.18. Motivations to reduce energy consumption at home, by country .. 101
3.19. Mean willingness-to-pay for renewable energy in 2011, OECD(11) 102
3.20. Willingness-to-pay for renewable energy, by country 103
3.21. Reasons for not wanting to pay any extra for renewable energy .. 104
3.22. Willingness-to-pay for renewable energy by level of trust in environmental information from government 105
3.23. Relationship between willingness-to-pay and income level 106
4.1. Reported use of land-based transportation by mode (2008) 116
4.2. Environmental concern clusters 118
4.3. Environmental concern and household location 119
4.4. Reported car ownership rates by country 119
4.5. Reported distribution of fuel types among vehicle owners 120
4.6. Reported cars owned, by household size and respondent's occupation 121
4.7. Reported number of cars owned per household, by income 122

4.8. Car ownership, air quality, and household location 122
4.9. Reasons for not owning a car, by environmental concern 123
4.10. Stated importance of factors in car purchases 124
4.11. Importance of different factors in car purchases, by level of environmental concern 125
4.12. Reported ownership of alternative-fuel vehicles by type and by country 126
4.13. Environmental attitudes and ownership of alternative-fuel vehicles 127
4.14. Willingness-to-pay for an electric car versus a conventional car, by country 127
4.15. Respondents NOT willing to pay more money for an electric vs. conventional car 128
4.16. Reasons for not wanting to pay extra for an electric car, by environmental concern cluster 129
4.17. Willingness-to-pay for electric car, by attitude towards costs of environmental policies 129
4.18. Reported car use, by country and by gender 130
4.19. Reported car use, by employment status and household size 131
4.20. Reported car use, by country and by residential area type 132
4.21. Reported car use, by access time to nearest public transport stop . . . 133
4.22. Car use, by country and by environmental concern cluster 133
4.23. Car use, by satisfaction with aspects of local environment 134
4.24. Reported assessment of measures to reduce car use 135
4.25. Reported frequency of transport modes for travelling to work . . . 136
4.26. Reported frequency of transport modes for shopping trips 136
4.27. Mean annual income by type of trip and most frequent mode . . . 137
4.28. Reported commuting modes, by travel time with fastest alternative 138
4.29. Distribution of commuting travel times, by mode 139
4.30. Main mode for commuting and access time to public transport stop 139
4.31. Main mode for shopping trips, by store type 140
4.32. Public transport and car mode shares for commuting, by activity level and by gender 140
4.33. Public transport mode shares for commuting, by country and by residential area type 141
4.34. Main mode for commuting, by environmental concern cluster . . . 142
4.35. Public transport share in commuting trips, by distance to public transport stop and by environmental concern cluster 142

4.36. Factors encouraging public transport use, by environmental concern cluster ... 143
4.37. Support of government actions to reduce CO_2 emissions, by country ... 144
5.1. Percentage of households facing a volumetric water charge, 2011 and 2008 ... 155
5.2. Investments in appliances and receipt of financial support to make these investments (in percentage) ... 156
5.3. Taking water efficiency into account when purchasing a washing machine or dishwasher ... 156
5.4. Frequency of undertaking water-saving behaviours ... 157
5.5. Spearman correlation between water-saving behaviours and unit pricing ... 158
5.6. Relationship between watering the garden in the coolest part of the day to save water and unit water charge ... 159
5.7. Relationship between collecting rainwater/recycling waste water and unit water charge ... 160
5.8. Relationship between plugging the sink when washing dishes and supporting/participating in an environmental organisation ... 161
5.9. Relationship between watering the garden in the coolest part of the day to save water and supporting/participating in an environmental organisation ... 162
5.10. Relationship between presence of children and index of water conservation behaviours ... 163
5.11. Correlation between water-saving investments and unit pricing ... 163
5.12. Relationship between adoption of low-volume or dual-flush toilets and unit water charges ... 164
5.13. Investment in low-flow toilets and ownership status for those facing unit water charges ... 165
5.14. Relationship between adoption of water tank to collect rainwater and supporting/participating in an environmental organisation ... 166
5.15. Relationship between taking water efficiency into account when purchasing appliances and respondents' recognition and use of water labels ... 167
5.16. Correlation between receiving water conservation grants and selected variables ... 168
5.17. Relationship between income and receipt of grants for different water-efficient devices ... 169
5.18. Relationship between investment in water-conservation devices and index of water-saving behaviour ... 169
5.19. Water-saving behaviour and investments, and demographic characteristics ... 170

5.20. Relationship between satisfaction with tap water and source of drinking water 171
5.21. Relationship between drinking tap water and satisfaction with its taste 172
5.22. Relationship between satisfaction with tap water and demographic characteristics 173
5.23. Relationship between drinking tap water and demographic characteristics 173
5.24. Importance level of factors that encourage people to reduce water use 176
6.1. Mean percentage expenditure for organic fruit and vegetables, by country 186
6.2. Organic fruit and vegetables: Comparing mean expenditure in the 2008 and 2011 surveys 187
6.3. Mean expenditure on organic fruit and vegetables and involvement in an environmental organisation 190
6.4. Mean expenditure for organic fruit and vegetables and seriousness of environmental issues 191
6.5. Mean expenditure for organic food and level of concern with environmental issues 192
6.6. Food waste by age group, OECD(11) 198
6.7. Percentage of food that is thrown away and rank of environmental concerns 199
6.8. Percentage of food that is thrown away and stated concern for selected environmental issues 199
6.9. Mean percentage expenditure and trust in manufacturers and retailers 200
6.10. Mean percentage expenditure on fruit and vegetables labelled as organic and national organic label recognition and trust 203
6.11. Median willingness-to-pay for fresh fruit and vegetables labelled as organic 205
6.12. Mean willingness-to-pay by age group and income quintile, OECD(11) 206
6.13. Mean willingness-to-pay and involvement in an environmental organisation 207
6.14. Mean willingness-to-pay and trust in manufacturers and retailers 208
6.15. Mean willingness-to-pay and label recognition and trust 209
7.1. Image displayed to respondents to help them estimate their household waste 223
7.2. Average weekly household waste per person, by country and household size 224

7.3. Average weekly household waste, by country and by income quintile 225
7.4. Average weekly household waste by rank of environmental issues 226
7.5. Average weekly reported waste generated and index of environmental attitudes 226
7.6. Households reporting having a pay-as-you-throw and other billing systems for mixed waste disposal, by country 227
7.7. Comparison of volume-based versus flat fee impacts on waste generation in Japan, Korea and Switzerland 228
7.8. Respondents not knowing how they are billed for their mixed waste according to length of residence 228
7.9. Respondents disposing of their hazardous waste with their mixed waste collection 229
7.10. Households disposing of their hazardous waste with mixed waste collection, by age group 230
7.11. Disposing of hazardous waste and perceived seriousness of waste-related environmental impacts 230
7.12. Average number of different materials separated 231
7.13. Waste separation and importance of environmental concerns relative to other global issues 232
7.14. Households using recycling services, by country and by service type 233
7.15. Households who stated having deposit-refund systems 233
7.16. Waste generation by availability of recycling services and by country 234
7.17. Household separation rates by type of waste, service availability, and by country 235
7.18. Factors motivating households to separate waste 237
7.19. Households who state not being informed about recycling services availability 238
7.20. Households' support for five different waste-reduction policies 239
7.21. Trust in sources of information about environmental impacts 240
8.1. Median willingness-to-pay for different environmental goods 249
8.2. Reasons for not being willing to pay extra 250
8.3. Reported effect on resource use 252
8.4. Factors that would encourage reduced water or energy consumption 252
8.5. Recognition of different labels 253
8.6. Trust and use of different labels 254

8.7. Commuting by public transport and willingness to compromise lifestyle ... 255
8.8. Recycling rates and willingness to compromise lifestyle ... 255
8.9. Respondents with high-tech environmental products by innovative technologies ... 256
8.10. Average (inverted) rank of issues in 2008 and 2011 surveys ... 257
8.11. Mean reported seriousness of environmental concerns in 2008 and 2011 surveys ... 258
8.12. Responses to selected attitudinal statements ... 259

Acknowledgements

This book is a product of the Working Party on Integrating Environmental and Economic Policies, a group under the OECD's Environmental Policy Committee (EPOC). The delegates provided valuable direction, comments and suggestions.

The OECD work on Environmental Policy and Household Behaviour benefited from the support of a number of contributors including: the Australian Department of the Environment and Water Resources; the Dutch Ministry of Economic Affairs, Agriculture and Innovation; Environment Canada; the French Ministry of Ecology, Sustainable Development and Energy (MEDDE); The Israeli Ministry of Environmental Protection; the Japanese Ministry of the Environment; the Korean Ministry of Environment; the Swedish Energy Agency (STEM) and the Swiss Federal Office for the Environment. The Secretariat gratefully acknowledges the financial contributions which allowed it to carry out this work.

The Secretariat is also grateful to the Advisory Committee which was set up to help inform the project and to ensure the political relevance of the outcome of this work. The Committee members from countries participating in the survey, OECD Directorates working in related areas, the International Energy Agency (IEA) and academy met regularly to provide guidance on the questionnaire design and the survey implementation, and to discuss the results.

The Secretariat developed the project and co-ordinated the work of several expert teams with extensive experience who prepared technical reports for the OECD which served as the main inputs to the five thematic chapters of this publication: Prof. Bengt Kriström (Umeå-SLU University, Sweden) – *Energy*; Prof. Kay W. Axhausen and Dr. Claude Weis (Swiss Federal Institute of Technology) – *Transport;* Prof. Quentin Grafton (The Australian National University) – *Water;* Dr. Katrin Millock (CNRS-Centre d'Economie de la Sorbonne and Paris School of Economics, France) and Dr. Céline Nauges (The University of Queensland, Australia) – *Organic Food;* Prof. Ofira Ayalon, Sharon Brody, and Prof. Mordechai Shechter (University of Haifa, Israel) – *Waste.*

This project is managed by Ysé Serret-Itzicsohn. This publication has been prepared by Zachary Brown, Nick Johnstone and Ysé Serret-Itzicsohn.

The assistance of Šárka Svobodová in the formatting of the manuscript is gratefully acknowledged.

List of acronyms

AFVs	Alternative-fuel vehicles
AUD	Australian dollar
CAD	Canadian dollar
CHF	Swiss franc
CLP	Chilean peso (the currency symbol is USD)
CO_2	Carbon dioxide
EEA	European Environment Agency
EPIC	Environmental Policy and Individual Behaviour Change
ELVs	End-of-life vehicles
EU	European Union
FITs	Feed-in tariffs
GHG	Greenhouse gas
IEA	International Energy Agency
ILS	Israeli new shekel
JPY	Japanese yen
KWh	Kilowatt hour
KRW	Korean won
LCA	Latent class analysis
LPG	Liquefied petroleum gas
MSC	Marine Stewardship Council
NGOs	Non-governmental organisations
PAYT	Pay-as-you-throw
PM	Particulate matter
PV	Photovoltaics
SEK	Swedish krona
VAT	Value-added tax
WTP	Willingness-to-pay

Executive summary

Personal behaviour and choices in daily life, from what we eat to how we get to work or heat our homes, have a significant effect on the environment. Their impacts are likely to intensify over the coming years without stronger and better-targeted policy efforts. How should governments respond? We need to intensify our efforts at developing growth strategies that promote and win support for more environmentally benign lifestyles and consumption patterns.

This publication – based on the Environmental Policy and Individual Behaviour Change (EPIC) survey, carried out in 2011 – helps governments to better understand household behaviour towards the environment in five key areas: energy use, water use, transport choices, food consumption, and waste generation and recycling. The second of its kind (the first was carried out in 2008), the survey collected information from more than 12 000 households across Australia, Canada, Chile, France, Israel, Japan, Korea, the Netherlands, Spain, Sweden and Switzerland. It also identifies policies that work to promote "greener" behaviour at the household level.

Based on the data, survey respondents can be grouped into three main categories when it comes to their environmental attitudes: *i)* the "environmentally motivated" who believe that sacrifices will be necessary to solve environmental problems; *ii)* the "environmental sceptics" who believe that environmental problems are often exaggerated; and *iii)* a group of "technological optimists" who believe that environmental problems are real, but that technological innovations are key to solving them.

For all these groups, governments need to show convincing evidence not only that changing behaviour is necessary to meet the challenge of scarce resources and climate change, but also that individual households' choices can make a difference, from recycling to choosing public transport. And where people are willing to change, governments need to have the policies ready to help them do it. They also need to take account of the gap between people's good intentions and actual behaviour.

Findings from the survey reinforce the need for the right economic incentives to influence people's decisions. Consistent with the 2008 survey, findings from the 2011 survey underline the significant role of environmental attitudes in shaping behaviours. People will not make an effort to go green if they do not believe there is any real benefit from doing so. In addition, in areas such as energy, water and personal transport, scaling up services and infrastructure is critical. You cannot forsake your car for public transport if there is none available that goes where you need to be, or switch to cleaner energy if there is no supply available. And for specific groups of households that cannot afford to take up more environmentally benign practices, providing grants may also be warranted. Spurring desirable behaviour change therefore requires a mix of instruments.

Key findings

- There is significant unmet household demand for electricity generated from renewable sources. Around 60% of respondents would be willing to pay extra for such electricity while 45% express an interest in having differentiated rates for renewable energy if this option were available to them.
- Most respondents in each country are engaged in some form of energy-saving behaviour. However, 40% of respondents report that they "occasionally" or "never" completely turn off appliances with stand-by functions. On average, higher-income households engage less frequently in energy-saving behaviours.
- Water charges based on the amount of water used increase households' efforts at water conservation, both in terms of investments and habitual behaviour.
- Governments play an important role in promoting household investments in energy efficiency. Households reported receiving government support for around 16% of the energy efficiency investments recorded in the survey.
- Energy efficiency labels also play a role in reducing electricity demand. Households who recognised energy efficiency labels for appliances spent on average 6% less on electricity than households who did not recognise these labels.
- There is a significant stated willingness to pay an additional price premium for the purchase of electric cars, although actual ownership remains very low. There is broad stated support for additional government investment in public transport infrastructure.
- Households' stated mean expenditure on organic fresh fruit and vegetables varies across countries and ranges from 13% to 35% of total expenditure on these products.

- There is wide variation across countries in the levels of recognition and trust in labels. For example, trust in the new European Union organic food label varies from 47% (Sweden) to 83% (Netherlands) among respondents who recognised it.
- Waste generation tends to be between 20% and 30% lower for households subject to pricing of waste by volume or weight. The two policy measures that respondents most strongly supported in terms of waste generation rates relate to waste prevention – encouraging retailers to use less packaging, and households to purchase products with less packaging.
- In all six countries involved in the two rounds of the survey, there was a significant increase in the percentage of respondents who felt that environmental issues should be dealt with primarily by future generations, although older people felt that it was up to them as the generation that created current problems.

Key recommendations

- Measures that increase consumer access to greener choices, such as investment in infrastructure (e.g. public transport or recycling services), are important complements to policies that make green choices cheaper.
- Need-based grants for water efficiency investments could provide an important means of improving water conservation.
- Households who rent rather than own their homes make fewer financial investments in water efficiency. Programmes for increasing water-saving investments among tenants could be a useful way to address this issue.
- Household demand for electricity does not depend on household income levels. This means that, without additional policy measures, higher energy prices are likely to have adverse welfare impacts on low-income households without significantly reducing consumption.
- Scaling up public information and educational campaigns is critical for raising household awareness of costs and charges (e.g. of waste collection or water consumption) and increasing understanding of climate change.

Greening Household Behaviour
Overview from the 2011 Survey

Chapter 1

The environmental policy context

by
Ysé Serret-Itzicsohn, Zachary Brown *and* Nick Johnstone*

This chapter reviews some of the main policies implemented by governments to influence household behaviour in five areas: energy use, water consumption, waste generation and recycling, food consumption and personal transport choices. It presents respondents' perceptions of policies in place and also provides an overview of measures which were actually adopted in the eleven countries when the OECD Survey on Environmental Policy and Individual Behaviour Change (EPIC) was implemented in 2011. The use of unit-based charging for environmental services is examined for waste collection, water use and electricity consumption. Differences across the countries surveyed are presented about charging systems in place, provision of grants to encourage households to invest in eco-friendly equipment, use of eco-labels and access to infrastructures such as collection services for recyclable materials and public transport. Keeping this broad picture in mind, and also some country-specific aspects, is essential when reviewing the data collected.

* OECD Secretariat, Environment Directorate. Nicholas Lancaster, Alessandro Giovannini, Anne Meldau and Amélie Rudloff from SciencesPo Paris provided inputs for the preparation of this chapter.

The statistical data for Israel are supplied by and under the responsibility of the relevant Israeli authorities. The use of such data by the OECD is without prejudice to the status of the Golan Heights, East Jerusalem and Israeli settlements in the West Bank under the terms of international law.

1. Environmental policies targeting household behaviour

The objective of the OECD EPIC Survey is to collect data which allow us to assess the effects of different types of policy measures on household decision making in five areas: energy use, water consumption, waste generation and recycling, food consumption and personal transport choices.

A broad variety of instruments can be used by governments to influence behaviour in these areas, including economic instruments (e.g. waste charges, grants for insulation), direct regulation (e.g. water use restrictions, technical standards of appliances), labelling and information campaigns (e.g. eco-labels), as well as the provision of environment-related public services (e.g. recycling schemes, public transport).

These policy measures provide different incentives for "environmentally responsive" consumer choices and behavioural responses, whether by changing relative prices of more and less environmentally damaging options, constraining or expanding the choices available to consumers, or providing information which helps them make more informed choices.

Economic instruments, such as environment-related taxes, have a direct effect on the relative prices of different goods and services. They promote consumption choices which reflect associated environmental impacts, even if consumers are not aware of these impacts directly. However, the relative efficiency of these instruments will depend very much on the extent to which taxes can target the environmental damage as closely as possible. In some cases, this will be impossible (or excessively costly in administrative terms) and the tax target will be a proxy for the underlying environmental objective.

In some sense, subsidies, like economic incentives to buy electric cars, will have a similar effect as environment-related taxes on relative prices, and thus will encourage a change to less polluting alternatives. There is, however, an important distinction. By subsidising the consumption of a less environmentally damaging product or input, the result will be increased consumption overall. The importance of this effect will depend on the relative price and income elasticities. Perhaps more importantly, it can be difficult to target subsidies efficiently, at the level of either the good (e.g. energy-efficient appliances) or the beneficiary of a programme (e.g. insulation programmes). Problems of moral hazard, free-riding and adverse selection can arise.

Direct regulation is certainly the most frequently used approach in OECD countries to influence the environmental impacts of household decisions. Often, regulation constrains the choices available to consumers. For example, standards on the energy efficiency or water efficiency of appliances remove more "wasteful" goods from the market. Direct regulation can be effective, and is often efficient. For example, bans on the disposal of some particularly noxious waste streams are likely to be necessary. More generally, there is a strong case to be made for the use of performance standards rather than technology-based standards. However, even with performance standards, consumers with different demand and market conditions are not able to trade off product attributes or behavioural choices in a manner that reflects their underlying preferences. This results in greater overall social cost.

Policy makers can also rely on information-based instruments such as eco-labels. This enables households to take more informed decisions with respect to both the private (e.g. financial cost) and public (e.g. environmental impacts) consequences of their choices. Information on fuel efficiency or food attributes is an example. More generally, public information campaigns will raise households' awareness on the environmental impacts of their consumption patterns. Assuming that there is an underlying demand for environmental quality, this will affect the choices made by households in the market. However, trust in the source of information is important, as well as ease of recognition and understanding of labels.

Finally, policy makers can increase households' access to goods or services that facilitate their ability to give up environmentally damaging practices and to adopt environmentally benign ones. This is important because many of the most environmentally significant sectors are "imperfect markets" for non-environmental reasons, thus necessitating government intervention. For example, since governments play a direct role as provider or regulator of transport, energy and water services, they have significant influence over the characteristics of such services.

More specifically, the broad range of policies targeting household behavioural adjustments in the five areas examined includes:

- For residential energy use: *energy taxes, energy efficiency labelling of appliances and buildings, grants to invest in energy-efficient equipment, minimum energy performance standards (MEPS) for appliances and lighting, provision of differentiated "green" energy…*
- For personal transport choices: *fuel taxes, congestion charges, fuel economy standards, subsidies for alternative-fuel vehicles, parking restrictions, emission standards, quality of public transport…*
- For residential water use: *pricing structure (fixed rate* vs. *increasing block tariff), grants for using water-efficient technologies, water efficiency labelling…*
- For organic food consumption: *product labelling, information campaigns on organic food products, organic standards…*

- For waste generation and recycling: *waste collection and management charges (flat fees vs. volume- or weight-based charges), deposit-refund systems, door-to-door* vs. *drop-off, recycling schemes, labelling schemes (e.g. recycled content)…*

Table 1.1 provides a summary of different policies applied in the five areas and examples of measures.

Table 1.1. **Examples of measures targeting household behaviour**

	Energy	Food[1]	Transport	Waste	Water
Economic instruments					
Taxes/charges	Electricity charges. CO_2 tax on fuel.		Fuel taxes, registration taxes. Charges for road usage.	Unit-based waste charges based on volume or weight. Advanced disposal fee.	Water charges.
Grants/subsidies	Grants for installation of solar panels. Free distribution of low-energy light bulbs.	Subsidising production.[1]	Reduced sales tax on alternative-fuelled vehicles.	Refund for recyclable bottles.	Reduced VAT for water-efficient appliances.
Direct regulation					
Performance/ technology standards	Minimum thermal efficiency standards for new dwellings.		Maximum CO_2 emissions for new cars. Maximum sulphur content in diesel.	Minimum recycled content standard.	Minimum water efficiency standard for dishwashers. Mandated use of dual-flush toilets in new buildings.
Bans/mandates	Mandated double-glazing of windows.		Mandated use of catalytic converters.	Bans on presence of toxics in certain products. Take-back requirements.	Temporary water restrictions.
Information-based measures					
Labels	Energy efficiency labels for appliances and buildings.	Organic food labelling.	Fuel consumption and CO_2 label for cars.	Label indicating manufactured from recycled materials.	Water efficiency label for washing machines.
Information campaigns and education programmes	Tips to save energy at home.	Information on environmental and health benefits of organic food.	Tips on fuel-efficient driving.	Tips on how to recycle and dispose of toxic waste.	Awareness campaigns on water scarcity.
Supply measures – e.g. provision of environment-related services	Differentiated tariff option according to time of use. Green energy option. Rolling-out of smart electricity meters.		Public transport. Provision of cycling paths. Bike-sharing or car-sharing systems (e.g. Velib in France). Charging facilities for electric vehicles.	Provision of waste collection and recycling services (door-to-door *vs.* drop-off containers).	Rolling-out of smart water meters.

1. In the case of organic agriculture, most policy measures are targeted on the supply side. The only measures which are targeted directly on households are labels and public information campaigns.

In the five thematic sections of the EPIC questionnaire (see Annex A), respondents were asked to indicate if they were subject to a number of different policy measures. This chapter presents respondents' perceptions of various policies in place in their country, such as the type of charging system for waste, the use of eco-labels or the availability of grants to encourage investment in eco-friendly goods. Moreover, it also provides an overview of selected measures which were actually implemented when the 2011 survey was conducted and which are expected to influence household behaviour in the areas examined. Keeping this broad picture in mind, but also some country-specific attributes, is key when reviewing the data collected.

2. The use of charges

As expected, there are differences across the countries surveyed in terms of the use of unit-based charging for environmental services. The greatest variation can be observed for waste charges. Korea,[1] Switzerland and Japan have the highest number of respondents reporting that they are charged for the collection of mixed waste in their primary residence according to volume or weight (see Figure 1.1).

As a matter of fact, a volume-based waste fee system has been applied in Korea since 1995. It is a unique example of a pay-as-you-throw (PAYT) unit pricing scheme introduced at the national level. The use of variable charging for Japanese households has also progressed over the past years and most municipalities have

Figure 1.1. **Percentage of respondents reporting being charged on a unit basis for different environmental services**

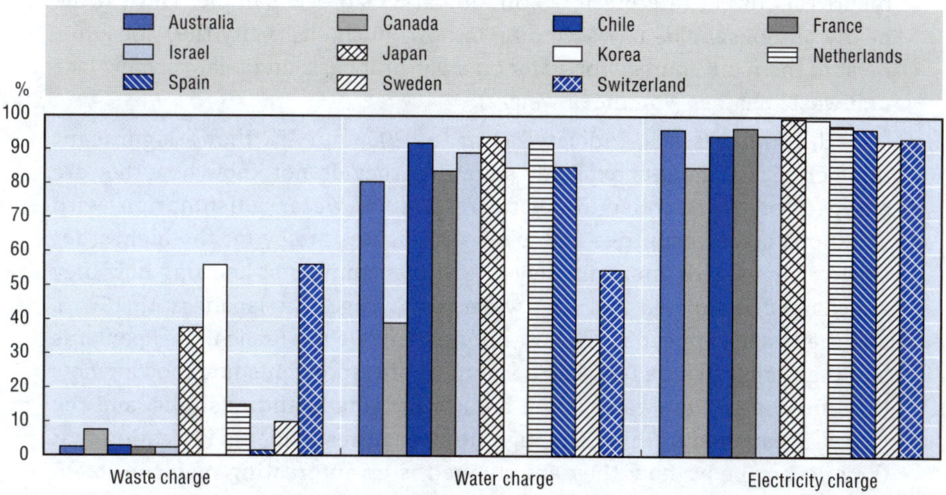

Note: Respondents reporting that they do not know how they are charged for these different environmental services are not included in the percentage. While replies are presented for all the countries surveyed, one should be cautious with country data when reported figures are below a certain percentage as this may reflect situations where a policy is just not in place (e.g. waste is not charged on a per-unit basis in Chile).

implemented volume-based schemes where waste is collected only when placed in special prepaid bags. A large percentage of the Swiss population is also charged according to the volume of waste disposed of and a system of "bag tax" was introduced with fines for not paying the disposal fee of up to CHF 10 000.

However, a high percentage of households do not pay for their waste-related services according to the amounts generated. While billing according to the frequency of waste collection is found in parts of the Netherlands and Sweden, in other countries, households predominantly pay a flat fee for waste collection management, such as a lump sum included in property taxes, charges or rent. It can be noted that, on average, one-third of the respondents report not being charged or not knowing how they are charged. This figure is the highest in Japan, followed by Israel, and the lowest in Switzerland.

There is greater use of volumetric water charging in general. Canada and Sweden, where water resources are abundant, have the lowest percentages. However, it is clear that, in practice, a higher percentage of households actually face volumetric water charges in these two countries relative to those who stated that they do. Part of the reason for this finding is that tariff structures are often complicated, with a mixture of fixed and unit charges, combining water, waste water and sewerage charges. This could lead to some confusion. Moreover, according to the Bloomberg New Energy Finance database of water tariffs, these two countries had relatively low tariffs in comparison with other countries in the sample.

The patterns of electricity charging are very similar, with around 90% of the households charged on a per-unit basis. As in the 2008 survey, Canadians report the lowest rate but it is still almost 85% of the sample. Thus, while nearly all households pay according to how much electricity they consume, most of them are charged for water on a per-unit basis and relatively few face unit waste charges (volume or weight).

The findings also indicate, as in the 2008 survey, that a significant number of respondents explicitly state that they do not know how they are charged for waste collection or for residential water consumption, with variations across countries and areas. The figures are by far the highest for waste charges with approximately 15% of the sample not knowing how they are charged, compared to 4% for water, on average. The Japanese, the Swiss and the Israelis appear as the best informed, while the French, the Spaniards and the Koreans come last. The percentage of respondents not knowing how they are charged for water is the highest in Canada and Australia, and the lowest in Japan and Chile. The high number of respondents indicating that they do not know how they are charged is an interesting result in itself, suggesting that an important task for governments and service providers is to raise household awareness. In order for a "price" to have an effect on resource use, users must be aware of the existence of such a price.

3. The use of grants to encourage households to invest in eco-friendly equipment or products

Governments also provide financial support as part of the policy mix to promote greener purchasing decisions. Support includes grants and subsidies for equipment such as roof-top solar panels, reduced value-added tax for water-efficient appliances or reduced sales taxes on alternative-fuel vehicles.

Grants towards the purchase of capital equipment

Respondents were asked if they had installed energy-saving equipment, bought solar panels and other renewable energy equipment or invested in devices to improve water efficiency at home. Figure 1.2 presents the percentage of those who have received financial support from the government to make these investments. Grants to invest in less polluting cars are examined below in a separate section.

Not surprisingly, households in water-scarce Australia and Israel report receiving support for investments in water-restrictor taps. Australia provides rebates and other incentives to encourage the installation of water-saving devices such as South Australia's H2OME rebate scheme. In Israel, the government supports country-wide free distribution of water-saving

Figure 1.2. **Percentage of respondents reporting having invested in eco-friendly equipment who received grants**

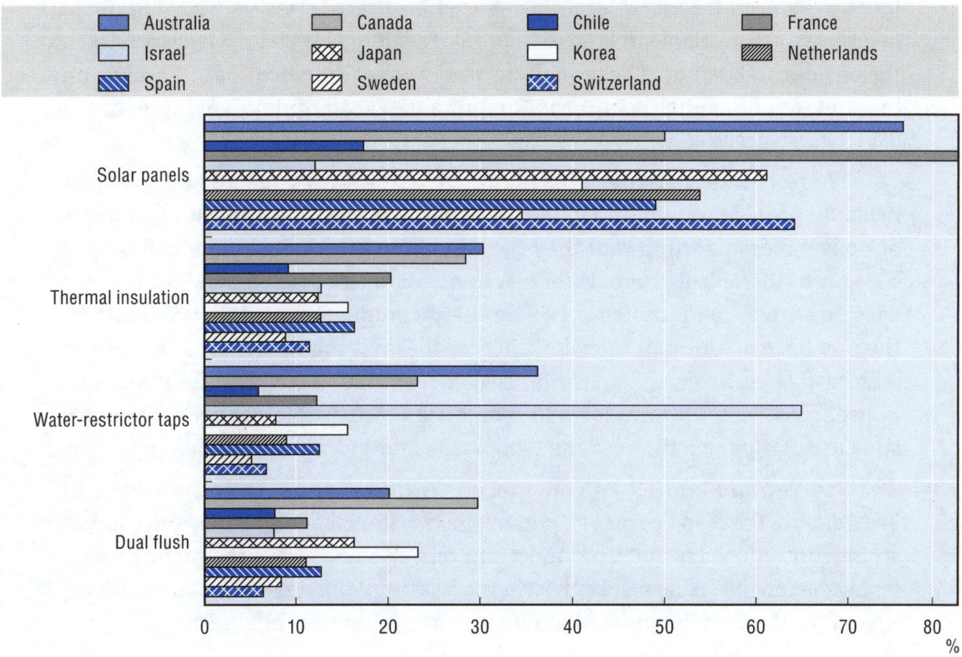

equipment.[2] Canadian households are the most likely to receive financial incentives for the installation of low-flow toilets. Grants for water-conservation improvements are offered under the *ecoEnergy Retrofit-Homes* programme and are distributed at the provincial level as well.

Australian and Canadian respondents are most likely to report having received grants for thermal insulation, with almost one-third of respondents who had undertaken such investments indicating that they had received support. French respondents are the third most likely to report having done so (20%). The results suggest that the Swedes are the least likely to do so. However, individuals in Sweden benefit from financial support in the form of tax deductions for renovation, maintenance, conversions and extensions (the ROT reductions) of homes. While the objective of the tax deduction is to stimulate demand for building services, and some of this may increase energy demand (i.e. extensions), support is received for a number of measures which reduce energy use.[3] In addition, in the period just before survey implementation, financial support was explicitly provided for investment in solar cells.

In Australia, a number of rebates promote energy-saving devices such as that which is available for the replacement of an electric water-heating system with an approved gas space heater. In Canada, the *ecoEnergy Retrofit-Homes* programme also helps home-owners to invest in energy-efficient upgrades. Insulation is part of the eligible retrofits together with upgrades in the heating and cooling system and the replacement of water-heating systems. Various incentives are offered by some provinces and territories as well. Financial incentives are available in France to promote energy efficiency investments in the residential sector. These include the interest-free eco-loan targeted on low-income households and the Sustainable Development saving account (*Livret développement durable*).

There is much greater cross-country variation as regards support for investment in solar panels. France and Australia have the highest percentage of households reporting that they have benefited from government financial support to install solar panels for electricity or water. The French government uses incentives to promote renewable energy production such as tax credit for the purchase of either photovoltaic (PV) or thermal solar panels, as well as the *interest-free eco-loan*. In Australia, the government provides assistance to households with upfront costs of small-scale renewable energy technology such as solar panels through the Renewable Energy Target scheme.

Switzerland and Japan come after, with 64% and 61% respectively. In Switzerland, the installation of a photovoltaic system is promoted by the support scheme introduced in 2009 and most cantons also encourage the production of renewable energy. A new subsidy scheme was implemented in Japan in 2009 to encourage the installation of solar PV systems in the residential sector.

In Canada, federal rebates for solar thermal systems exist, as well as incentives at the level of province and territories. A direct capital subsidy for PV systems has been available in Sweden since 2009 and is expected to last until 2016. And between 2009 and the end of 2011, the Swedish government also introduced a support scheme for installing a solar heating system, with the amount of support based on the quantity of heat produced by the system. Other initiatives to encourage the installation of renewable energy technology in the countries surveyed include low-interest rates for owner-occupiers to install solar panels in the Netherlands or grants provided to Spanish households investing in solar thermal systems.

While Israel does not rank high in Figure 1.2, it should be noted that it is by far the country where the largest number of solar panels are installed in the whole sample, followed by Australia.[4] The very high rate of solar panel installation is associated with the regulations in place since the mid-1970s requiring most buildings to be equipped with solar water heaters.

Renewable energy generation and feed-in tariffs targeted at households

In addition to grants encouraging investment in capital equipment to produce renewable energy, households in some countries can benefit from feed-in tariffs (FITs) for the electricity they generate and feed into the grid. The possibility to sell back the surplus of electricity they generate to energy suppliers provides additional incentives to invest in low-carbon electricity technologies (e.g. solar photovoltaic, wind) in the residential sector.

While FITs are widely used, they are not always available to small-scale producers. However, in some cases, tariff rates can be set at a higher level for small producers specifically to encourage them. In seven of the countries surveyed, households can benefit from FITs: Australia, Canada, France, Israel, Japan, Spain and Switzerland. Other OECD countries with FITs accessible to the residential sector include Germany, the United Kingdom[5] and the Unites States.

In Australia, the Grid Buy Back Rate in the Northern Territory is an example of scheme whereby home-owners, landlords and tenants, are encouraged to install solar photovoltaic panels by being paid for each unit of electricity that they generate and that they export back into the grid.[6] In Canada, the FIT programme in place in Ontario since 2009 for micro-scale systems (< 10 kilowatt) allows households with solar panels to connect to the grid and be paid guaranteed electricity tariffs.

French residential dwellings can also resell electricity from solar photovoltaic panels since 2006, benefiting from guaranteed purchase prices. These were much greater than the market price of electricity paid through meters until 2010. The Israeli Public Utility Authority approved a feed-in tariff for solar power in 2008 with a 15 kWp maximum for residential installations. FITs were also set up for electricity generated from small-scale residential wind production, in 2009.

In Japan, a new feed-in-tariff started in 2012 and the system allows small-scale systems (< 10 kW) to receive the feed-in tariff when selling excess electricity back into the grid. In Switzerland, a feed-in tariff system was introduced in 2008 and Swiss home-owners can also receive preferential tariffs for power produced. Finally, as part of Spain's FIT legislation of 2008, a guaranteed tariff was available for small-scale photovoltaic units (< 5 kW), aimed at promoting the installation of rooftop solar panels However, in response to the current financial situation, some countries have revisited and cut back on the tariffs proposed, even retroactively (as Spain did).

Grants to invest in less polluting cars

In addition to looking at grants received to invest in water- and energy-saving equipment or renewable energy, the survey examined subsidies to invest in alternative-fuel vehicles. A new question was introduced in the 2011 EPIC questionnaire (see Question 51 in Annex A) in which respondents who stated that they had bought a car in the last five years were requested to indicate if they had paid less for it, either because they had returned an old car as part of a government vehicle scrappage programme or because they had benefited from financial incentives for purchasing a fuel-efficient and/or a low-polluting vehicle (reduced tax/registration fee or provision of rebate). Figure 1.3 presents a summary of the responses.

Figure 1.3. **Car buyers who reported benefiting from a scrapping scheme and/or financial incentive to buy a "green" car over the past five years**

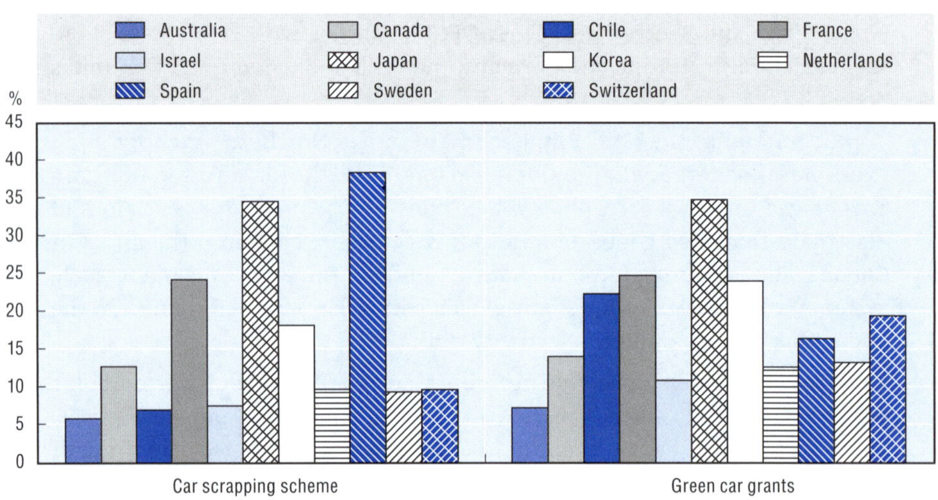

Spain has the highest percentage of respondents reporting that they have benefited from incentives as part of a scrapping scheme, with nearly 40%, followed by Japan and France. These countries, like many others, have introduced car-scrapping schemes where car owners received financial support to trade their old cars for more efficient ones. The programmes differ in their features, such as the amount provided and the eligibility criteria. Under the *Plan Vive* started in Spain in 2008, an interest-free loan of up to EUR 10 000 was granted for a period of five years for the purchase of a car if the scrapped car was at least 15-years old under certain conditions.[7] A new scheme was launched in 2009 (*Plan 2000E*) where the government provides support of EUR 500 per car, conditional on the car meeting certain CO_2 emissions criteria and the car manufacturers adding another EUR 1 000 per car. An extra EUR 500 is provided by some regions like Navarre or Valencia. A small percentage of respondents in some countries for which there is no official car scrapping scheme (such as Chile) actually report the existence of such a scheme. This may be due to the existence of schemes of private dealers in which consumers receive some compensation for older vehicles when purchasing a new one.

In France, the scrappage incentive scheme implemented alongside the *bonus-malus programme* provides an additional bonus of EUR 300 (*super-bonus*) if a vehicle older than 15 years is scrapped when purchasing a new car. In Spain, the scheme requires new cars to meet minimum emission standards. In 2009, Japan also introduced a scrapping scheme, which provided purchasing rebates to consumers trading in cars older than 12 years for fuel-efficient cars, according to environmental performance criteria established by the government.[8]

A programme was also initiated in Israel in 2010, whereby owners of vehicles at least 20 years old are eligible to receive NIS 3 000 (approximately USD 825) when returning their vehicles to authorised scrapping sites. Respondents from Israel were asked about their knowledge of this new scrapping scheme. More than 6 Israelis out of 10 were aware of that programme when the survey was implemented and 20% reported that they had already benefited or intended to benefit from it in the future. The fact the survey took place shortly after the measure was introduced needs to be taken into account when interpreting these results.

Other programmes in place in the countries surveyed include a tax break offered to Koreans replacing a car older than 8 years with a new one and a premium provided in the Netherlands which varied with the category of the vehicle scrapped (between EUR 750 and EUR 1 000). Sweden does not currently have a specific programme targeted on end-of-life vehicles (ELVs) returned for dismantling.[9] The scrappage programme was abandoned in 2007.

Scrappage programmes generally contribute to a slight class shift where medium-sized vehicles are replaced with lighter ones and to steering demand for more fuel-efficient cars, depending on their specific features. The financial cost associated with these incentive schemes needs however to be taken into consideration when assessing the efficiency of different policy alternatives.

As the review indicates, national scrappage programmes often appear to be combined with incentives to replace an old car by a more fuel-efficient one and/or to specifically support the take-up of alternative-fuel vehicles. A different approach is adopted in the Canadian initiative *Retire you ride*, implemented between 2009 and 2011, whereby environment-friendly transport options are offered as incentives, including public transit passes, membership in car-sharing programmes and discounts on bicycles.

The question introduced in the 2011 EPIC Survey (Question 51 in the questionnaire, see Annex A) also allows for a better understanding of different support schemes used to stimulate the purchase of fuel-efficient or green cars. Among those who report that they have purchased such vehicles in the last five years, Japanese respondents are the most likely to state they receive financial support (35%), followed by French, Korean and Chilean respondents (approximately 1 in 4).

These findings reflect the importance of the Japanese government's *Green Vehicle Purchasing Promotion Measure* which has been in effect since April 2009. A number of other incentives designed to accelerate the wider development of environment-friendly vehicles have been used in Japan over time.[10] These include reductions[11] on the car purchase tax, ownership tax and the motor vehicle tonnage tax, with tax exemptions for alternative vehicles such as electric and hybrid vehicles. In France, the tax differentiation scheme applied to new car sales since 2007 (*bonus-malus programme*) also provides a financial reward to the purchase of environment-friendly new cars. The bonus can reach EUR 5 000 for vehicles emitting 50 grams of CO_2 per km or less, like electric cars. Since January 2012, the Korean government also provides financial incentives for cars with low CO_2 emissions through tax exemptions on new electric vehicles. In Chile, financial support was provided between 2008 and 2010 to promote the purchase of hybrid vehicles where owners were refunded the cost of annual registration fees.

It can be noted in Figure 1.3 that less than 10% of Australian respondents report that they received an incentive to purchase a fuel-efficient or a low-emission vehicle. However, an incentive scheme to purchase new liquefied petroleum gas (LPG) vehicles or to convert petrol and diesel vehicles to LPG has existed since 2006.[12] A "cleaner car rebate" announced in 2010 was cancelled in 2011. The scheme was meant to provide a rebate of AUD 2 000 towards the purchase of a qualifying fuel-efficient vehicle when a vehicle manufactured before 1995 was scrapped.

Countries like Canada, Israel, the Netherlands, Sweden and Switzerland have an intermediate position in Figure 1.3. A number of incentives have been introduced by governments to encourage energy-efficient and alternative-fuel vehicles in these countries. The Canadian *ecoAuto Rebate Program*, implemented until 2009, offered rebates for the purchase of hybrid electric and highly energy-efficient vehicles. A variety of support programmes are available at the provincial and territorial levels for the purchase of hybrid vehicles. These include rebates on provincial sales tax or on the pre-tax car price. In Switzerland, financial incentives exist at the level of cantons in favour of energy-efficient vehicles or low CO_2 emissions such as motor vehicles rebates or reduced parking fees.[13]

In addition, the passenger car registration tax in Spain is linked to CO_2 emissions since 2008, with the least polluting vehicle being taxed at the lowest rates and vehicles with emissions under or up to 120g CO_2/km being exempted. A package of measures to promote electric vehicles was also approved in May 2011 as part of the *Electric Action Plan for 2010-12 (MOVELE)*. In Sweden, exemption from the annual road tax exists since 2009 for electric vehicles, vehicles using biofuels and fuel-efficient vehicles, for a period of five years upon first registration. In addition, rebates were provided for the purchase of environment-friendly vehicles under the *Eco Car Subsidy* in place between 2007 and 2009; a new programme started in January 2012 to provide financial support for the purchase of electric cars and other cars with ultra-low carbon emissions (below 50g CO_2/km).

Moreover, buyers of alternative-fuel vehicles benefit from strong government support in the Netherlands. A 2009 plan set an objective of one million electric vehicles on the roads in 2020. Since 2006 the government has provided a reduction on the vehicle registration tax depending on the amount of CO_2 emissions. Finally, in Israel, since 2008 vehicle taxation is linked to the pollution level emitted from the vehicle, and the government recently introduced economic incentives to encourage the purchase of environment-friendly cars.

Using positive financial incentives to steer demand for more fuel-efficient cars and green cars can have significant fiscal implications relative to taxing the most polluting vehicles. These need to be taken into account when designing a scheme and evaluating its environmental benefits. This can be illustrated by the French *bonus-malus* system which was designed to have a neutral effect on the state budget, but exceeded the government expectations and ended up costing more than EUR 200 millions in 2008.[14]

Direct regulation is another instrument in the policy mix available to governments to introduce more fuel-efficient cars to the market and remove the most polluting ones. EU regulations, for example, require that the average emissions of all newly registered passenger cars of each manufacturer should not be higher than 130 grams CO_2/km by 2015.[15] The Australian government plans to implement mandatory CO_2 emissions standards for all new light vehicles sold, starting in 2015.[16] Korea also announced a new and more stringent fuel economy standard for passenger cars, as part of the national *Green Growth Strategy*, to be phased in from 2012 and then fully implemented in 2015. However, since the point of incidence of such policies is not at the level of the consumers, these may not even be aware of their existence, the survey did not seek information.

4. The use of eco-labels

A wide range of eco-labels are used in the eleven countries surveyed. Table 1.2 presents an overall picture of the labels providing consumers with information related to the different areas covered in the EPIC Survey.

Table 1.2. **Use of eco-labels in the countries surveyed**

	Energy-efficient appliances	Energy-efficient buildings	Energy-efficient cars	Water efficiency	Organic food	Animal welfare	Waste-related
Australia	✓	✓	✓	✓	✓	✓	
Canada	✓	✓	✓		✓		
Chile	✓	✓	✓		✓		
France	✓	✓	✓		✓		
Israel	✓		✓	✓	✓		✓
Japan	✓	✓	✓		✓		✓
Korea	✓	✓	✓		✓		✓
Netherlands	✓	✓	✓	✓	✓	✓	
Spain	✓	✓	✓		✓		
Sweden	✓	✓			✓	✓	
Switzerland	✓	✓	✓		✓	✓	

Note: Chile's national fuel-efficient vehicle labelling scheme is under development. There are no car efficiency labels as such in Sweden but consumers have access to information on fuel consumption and carbon emissions for new passenger cars.

Labels providing information on the energy efficiency of appliances and organic food labels are the most common, followed by car labels and buildings energy efficiency labels. Water efficiency comes after, with only three countries using a specific label. A few countries also use animal welfare labels or waste-related labelling.

In each of the eleven countries surveyed, respondents were shown a selection of some of the most prominent eco-labels used in their own country. Labels/logos relating to the five thematic areas were displayed graphically to respondents, as well as more general eco-labels at the national or regional level (e.g. EU environment logo). Selected symbols which are used internationally such as the Marine Stewardship Council and EnergyStar were also presented. Figure 1.4 shows the level of recognition of different eco-labels.

Figure 1.4. **Recognition of different eco-labels**

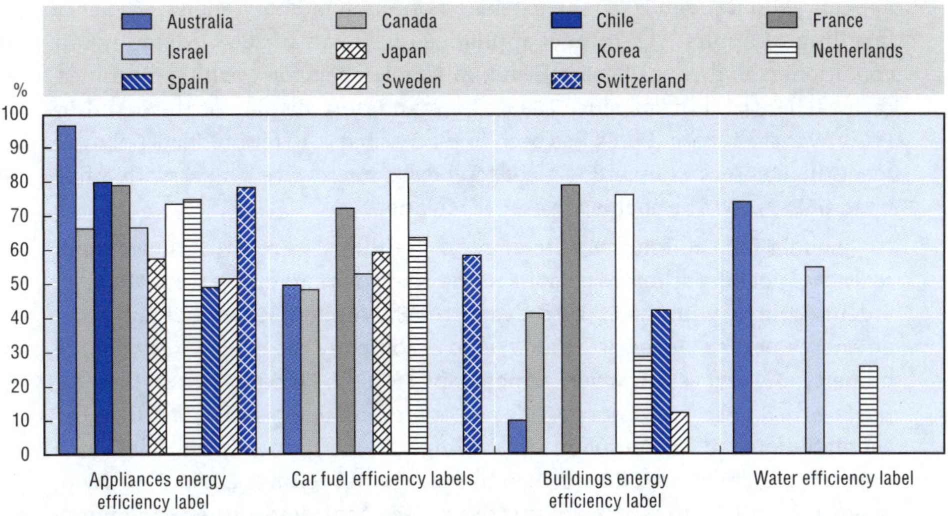

Note: The missing values correspond to two different cases: i) the label did not exist in the country when the survey was implemented in early 2012 (e.g. only three countries using water efficiency labels; in Chile, car efficiency label to be introduced in late 2012 and a voluntary process that determines the energy performance of new homes initiated mid-2012; buildings energy efficiency not available in Israel); ii) a label exists but was not displayed in the survey (e.g. car fuel efficiency label in Spain or labelling scheme for the energy efficiency of buildings in Japan and Switzerland).

Generally, as in the previous round of the survey, respondents were more likely to recognise energy efficiency labels for appliances. The results confirm that Australia has the highest percentage with 96%, followed by Chile, France and Switzerland. Korean and French respondents were the most likely to recognise car fuel efficiency labels (70% and 80%). The same pattern can be observed for the recognition of the buildings energy efficiency labels displayed in the questionnaire. Australians and Swedes are the least likely to recognise buildings efficiency labels. For water efficiency labels, the level of recognition in Australia is above 70% and Israel is second with more than one out of two respondents.

Energy efficiency labelling for appliances and buildings

Labels on the energy-efficiency of appliances are very widespread and were among the first to be introduced. Labelling of the estimated energy consumption of appliances is required in OECD/EU member states as part of the implementation of EU Directive 2010/30/EC.[17] Examples of schemes in other OECD countries range from the mandatory Canadian *EnerGuide* label for major electrical household appliances and room air conditioners to the Korean *Energy Efficiency Label* introduced in 1992 for energy-intensive appliances such as refrigerators, air conditioners and washing machines. Selected recent initiatives include: the revision in 2010 of the Australian *Energy Rating Labelling Scheme* launched in 1986 in which the star rating gives a comparative assessment of the appliance's energy efficiency; the introduction of an energy efficiency label for appliances in Chile in 2007; the extension of the mandatory energy efficiency labelling of appliances to new appliances in Japan in 2008 (including air conditioners and space heaters); and, in Israel, labels for washing machines, dishwashers and ovens since 2009. The mandatory display of the stand-*by Power Warning Label* in Korea has been in place since 2010 for products failing to meet the stand-by standard (e.g. set-top boxes, microwave ovens), with a fine up to USD 5 000 to be charged in case of violation.

Labels for the energy efficiency of buildings were introduced more recently. Labelling schemes exist in all the countries surveyed except Israel. In order to comply with the EU Directive on the Energy Performance of Buildings; an energy performance certificate has to accompany the sale of a residential property and real estate advertisement since 2011 in EU countries. In Canada, the *EnerGuide label for houses* provides an energy efficiency rating based on the estimated energy consumption. A labelling scheme for newly built detached houses has been introduced in Japan since September 2009. Finally, in May 2012 Chile initiated a voluntary process that determines the energy performance of new homes.

Car labelling

The use of labels to help consumers select vehicles with low fuel consumption is very common in the countries surveyed. In addition to displaying information on fuel consumption, certain labels include indicators of the vehicle's CO_2 emissions in g/km or other measures. Comparative labels in which relative emission performances between vehicles are provided exist in the Netherlands, Spain and Switzerland.[18]

Car labelling is mandatory in EU member states, as a result of the transposition of the 1999 directive[19] and related acts.[20] The introduction of energy efficiency classes using colours is an attempt to facilitate the harmonisation of labels. The French labelling system, *Etiquette énergie/CO_2*, is mandatory since 2006 for each new vehicle sold. It provides information about the car fuel consumption and its CO_2 emissions in g/km, with a colour rating system. In the Netherlands and Spain, a car label informing consumers about fuel consumption and CO_2 emissions was introduced in 2002 to comply with EU directives. The situation is somewhat different in Sweden where there is no car efficiency label as such at the national level. However, general guidelines handled by the Swedish Consumer Agency, which are mandatory since 2010, ensure that information on fuel consumption and carbon emissions for new passenger cars is made available to consumers.

Examples of initiatives in other OECD countries include car labels in Australia, Canada, Japan, Korea and Israel, as well as the recent Chilean labelling scheme. Australia introduced the obligation to display a fuel consumption label on cars in 2001. Since 2009, the new version of the label includes additional information on CO_2 emission values, as well as on urban and extra-urban fuel consumption. Canada introduced the *ecoEnergy* label for personal vehicles in 1999. It provides information on a vehicle's city and highway *fuel consumption rating*, and the estimated cost of fuelling the vehicle each year. Japan introduced a fuel efficiency labelling system in 2004, to promote public awareness of energy-efficient vehicles that achieved the *Top Runner* standards. The label shows fuel economy performance and indicates either the status of "fully compliant" or "plus 5% of the fuel economy standard". The system was revised, in 2012, to allow the identification of vehicles that are 10% or 20% more fuel-efficient.

Korea has a label for passenger cars since 2008 which displays fuel efficiency and carbon emissions,[21] and Switzerland uses an energylabel for passenger cars since 2002 to enhance transparency for consumers selecting or buying a new vehicle. The Swiss *energyEtikette* for passenger cars ranks passenger vehicles from "A/Green" to "G/Red" according to their fuel consumption in litres/100km relative to the car's weight. Estimated CO_2 emissions are noted. The revised fuel efficiency labelling for passenger vehicles is compulsory as of 1 January 2012 with a stronger emphasis placed on CO_2 emissions. In Israel, car fuel efficiency labelling is more recent. It is mandatory since 2009 and has to display information on fuel consumption and emissions of greenhouse gases. Finally, Chile launched in 2013 a mandatory national fuel economy labelling system to provide consumers with information on vehicles' CO_2 emissions and fuel efficiency.

Water efficiency labels

While car efficiency labels and energy efficiency labels are very common in the countries surveyed, only three of them use specific water efficiency labels to help consumers reduce their consumption by providing information on water-efficient appliances. Such labels were introduced in Australia and Israel, two countries with acute problems of water scarcity. In Australia, a mandatory national *Water Efficiency Labelling and Standards Scheme* (WELS) was set up in 2006, replacing the previous label. Products covered by the scheme include showers, taps, toilet equipment, dishwashers and washing machines. The star rating indicates water use and the scheme also contributes to water conservation by setting minimum water efficiency standards for selected products, currently toilet equipment and washing machines. The WELS star ratings are also used by state, territory and local governments to provide criteria for rebate eligibility.

The *Blue Label* is issued by the Israeli Water Authority and displayed on products such as taps, showerhead fittings and devices to reduce the amount of water used to flush toilets. It guarantees consumers that the product meets water-saving standards. The Netherlands uses a specific water efficiency logo (*laag verbruik*) which has been developed by an independent certification organisation.

Waste-related labels

Waste-related labels could in theory make reference to the recyclability of the product or its recycled content. Among the countries surveyed, such labels are rarely used. Japan and Korea appear as exceptions, as well as Israel where logos indicate the recyclability of materials (e.g. aluminium).[22]

Other types of waste-related labels are sometimes displayed on products, with no direct link to their own characteristics with respect to recyclability or recycled content. The French *Eco Emballages* stamp is an example.[23] A new question was introduced concerning labelling in the 2011 survey where respondents were asked how useful to them would be different types of information. One was on whether the product could be recycled (Q35, see Annex A). Findings suggest that Canadians, Chileans and Spaniards would find this information the most useful and Japanese and Dutch the least useful.

Food labelling

Food certification schemes have also been widely introduced as customers demand for quality has increased over time, in particular as a result of several food-related crises. Organic food labelling is the most widespread and labels exist in all of the eleven countries surveyed. The EU organic logo was redesigned in 2010 and it is compulsory on all packaged organic foods since July 2012, after a two-year transition period. In some EU member states, the EU label coexists with a national logo, in Spain for instance, or with private logos like in Sweden (e.g. KRAV[24] and the Demeter logos).

Some of the schemes introduced by non-EU member states include the new organic food label introduced by the government of Canada in 2009 which requires that domestic and imported products seeking organic certification meet Canada's standards. The Japanese organic label was launched in 2001 by the Ministry of Agriculture, Forestry and Fisheries under the *Japanese Agricultural Standard* (JAS) umbrella. It was extended in 2005. The Korean organic certification and labelling programme was revised in 2010 and extended to processed food products. Full implementation of the programme is scheduled for January 2013. In Australia, non-profit groups are privately certifying products asorganic such as the *Australian Certified Organic* (ACO) label and the NASAA *Certified Organic Label*. Organic food is also labelled in Chile and Israel.

In addition, in a few countries, a label provides consumers with information on animal welfare for products of animal origin. Australia has a *RSPCA Approved Farming* labelling scheme. Private certification schemes focusing only on animal welfare also exist in the Netherlands with the *Better Level Kenmerk* and in Switzerland. Some food labels may also cover several aspects, including animal welfare. Focusing on new developments, the adoption of a report by the European Commission outlines a series of options for animal welfare labelling. Korea introduced animal welfare labelling in 2012. Eggs were the first products concerned but the coverage of the programme is to be gradually expanded.

In a new question added to the 2011 EPIC questionnaire (Q80 in Annex A), respondents were asked whether information on the respect of animal welfare would be useful to them. The Swiss, who already have a label to help them identify welfare-friendly products, were the most numerous to rank it high while the Japanese and the Koreans were the least numerous to consider labels providing this type of information as useful.

5. Availability of environment-related services

As noted in the 2008 survey, supply-side measures which affect infrastructure provision and service access can usefully complement demand-side measures.

Access to infrastructures and services

Access to environment-related infrastructure and services in the eleven countries is first examined by checking the availability of door-to-door paper collection and access to the public services. Differences exist across countries as reflected in Figure 1.5.

Figure 1.5. **Reported access to selected environment-related infrastructures**

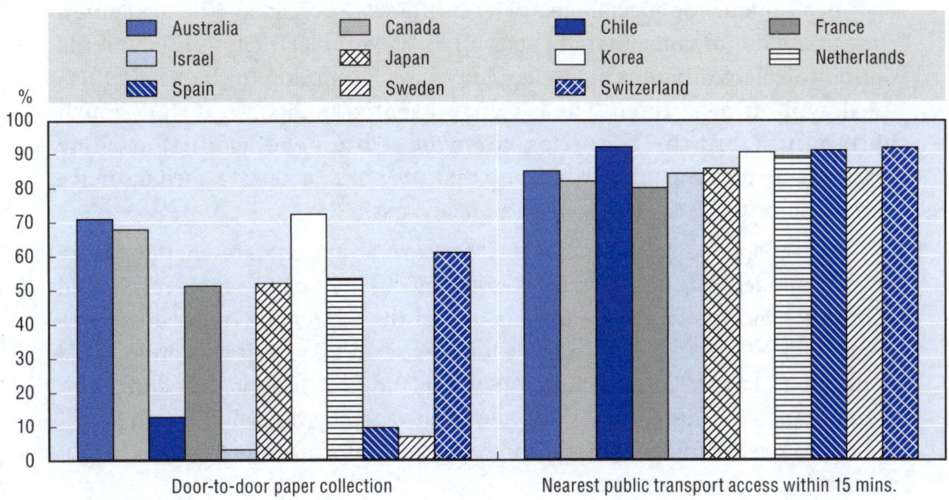

Door-to-door collection services for paper and cardboard are most common in Australia, Canada and Korea, followed by Switzerland. Just over one respondent in two also reports benefiting from this service in France, Japan and the Netherlands. Drop-off centres are more widely used in Israel, Spain and Sweden. However, it is important to note that the percentage of households with recycling services differs for other materials (see Chapter 7 on waste).

Providing transport-related services facilitates the adoption by households of greener travel practices. Respondents were asked how far they were from the most convenient public transport stop and how important was the provision of improved public transport in encouraging them to drive less. Figure 1.5 indicates that a very high percentage of households reported that they live within 15 minutes from a public transport stop convenient for their daily commuting, with little variation across the countries surveyed. Respondents were also asked if improved public transport was a very important factor in encouraging them to drive less; Koreans, Israelis and Chileans appear as the most likely to rank this factor as very important while the Dutch, the Canadians and the Australians ranked it the lowest.

Besides public transport availability and accessibility, infrastructures to promote alternative modes of transport include the construction of cycling paths, the development of bike-sharing and car-sharing systems in cities (i.e. the *velib* and the *autolib* programmes in France), as well as the deployment of electric vehicle charging stations where vehicles can plug in to an electrical source to recharge batteries.

Access to energy-related services

Two options for special electricity tariffs were considered in the survey: the possibility for consumers to sign up to "green" electricity tariffs and the option to select differentiated electricity rates according to the time of use (peak and off-peak time). The results presented in Figure 1.6 show much variation across the countries surveyed. Note once again that some respondents may mistakenly indicate that they have access to such a service even it is not available. The converse is also possible.

As a response to the increasing concern about climate change, some power suppliers are now offering a green electricity tariff to their residential customers. Access to these tariffs is one of the two energy-related services presented in Figure 1.6. Only a few countries seem to provide this option on a widespread basis. These include Australia, the Netherlands, Sweden[25] and Switzerland. The development of a system whereby households can select energy sources is under consideration in other countries, like Japan as part of the Electricity System Reform.

Figure 1.6. **Reported provision of environment-related electricity service options**

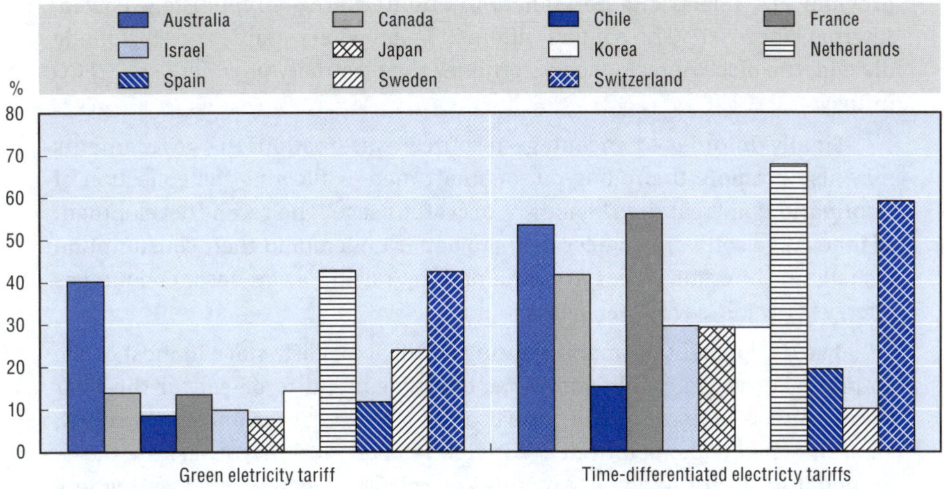

Australian households can selectthe amount of green power they want to buy from their energy provider. The *Green Power* programme began in New South Wales, in 1997, and was extended to other states/territories in 2000. In Switzerland, customers can also choose from different energy tariffs and electricity labelling is used to help them take informed decisions regarding the electricity product they buy. In Sweden, electricity suppliers offer many different contracts with 100% renewable energy sources to customers.[26] Dutch households also have the possibility to choose to buy "green electricity" from their electricity provider and its price is similar to conventional electricity as a result of tax incentives. Interestingly, the Netherlands is also the country with the highest percentage of respondents expressing no interest in having access to this service, while Chileans, Koreans and Spaniards, on the contrary, appear as the most inclined to take special action to buy green energy.

In addition, households have the possibility to select time-differentiated electricity tariffs on their bills to reduce peak loads and to shift demand to off-peak periods (e.g. night). Such rates have been available to Dutch consumers for a long time. Day and night tariffs are also applied in France, and in Switzerland where most households typically operate their boilers at night. Japanese households can select differentiated electricity tariffs according to time of use, as well as residents of Ontario in Canada. This service is also available to Spanish households since 2011.

The ease with which residential consumers' can change service provider to benefit from energy-related services, such as green tariffs or differentiated tariffs, varies significantly across the countries surveyed. Australian and Dutch households, for instance, are provided with a wide choice of energy providers and French households can switch from the main electricity provider EDF (Électricité de France) to an alternative supplier (e.g. Poweo, Alterna) since 2007. The context differs in Japan where utilities are regionally divided; the place of residence determines the electricity provider (e.g. TEPCO in Tokyo).

Finally, in order to encourage resource conservation, the governments may also promote the rolling-out of smart meters allowing the collection of information on "real time" energy[27] or water usage. The recent development of innovative software enables users to better understand their consumption visually in different places at home and, by raising awareness, encourages energy- or water-saving behaviour.

The deployment of smart metering is still at an early stage in most of the countries surveyed, while ambitious objectives tend to be set for the near future. Australia and Canada appear among the most advanced. Currently, smart meters are being introduced in Australia on a voluntary basis, with the exception of Victoria where a mandatory roll-out began in 2009, and where support for the widespread adoption of smart metering technology is growing. In Canada, initiatives to roll out smart meters exist at the provincial level. The government of Ontario was one of the first to have installed smart meters throughout the province by 2010 and these are to be deployed to all customers in British Columbia by the end of 2012.

Sweden was one of the first countries to roll out smart metering in the EU and the parliament decided in 2003 that all electricity meters should be read monthly by July 2009. In the Netherlands, the deployment of smart meters at the national level began in 2012. After facing strong opposition from consumers' organisations,the Dutch government opted for a two-phase voluntary roll-out of smart meters in the residential sector instead of adopting a compulsory approach.

In the survey, respondents were asked if they would be interested in having smart electricity meters installed and the Dutch appear as the least interested while the Chileans and the Israelis showed the greatest unmet demand.

Pilot projects with smarts meters in the residential sector are supported by the government in co-operation with utilities in Switzerland. In France, demonstration projects are also currently being conducted in different regions, before the national deployment phase involving the replacement of 35 million meters between 2013 and 2018. Smart meter initiatives are gaining momentum in Japan as well, where the government has been encouraging electricity companies to introduce smart electricity meters with ambitious objectives for 2016. Finally, the objective of the Spanish government is to install smart meters for all consumers with less than 15 kW by 2018.

6. Household attitudes towards environmental policies

There is variation in household exposure to different types of policies – from charges to grants and information-based instruments – across the surveyed countries. However, a significant issue addressed in the thematic chapters of this book focuses on the attitudes of individual consumers and households, and how these attitudes can interact with exposure to different policies to produce socially beneficial behavioural changes. This introductory chapter, as a bridge to the rest of the analysis, thus concludes with a presentation of some of the survey data relating to households' support for the different policies described above. The focus here is on two specific sets of questions relating to *i)* household support for different policies to address vehicle CO_2 emissions and *ii)* support for different policies to reduce household waste generation. Figures 1.7 and 1.8 summarise the data for these two sets of issues.

Figure 1.7. **Support for different policies to address vehicle CO_2 emissions**

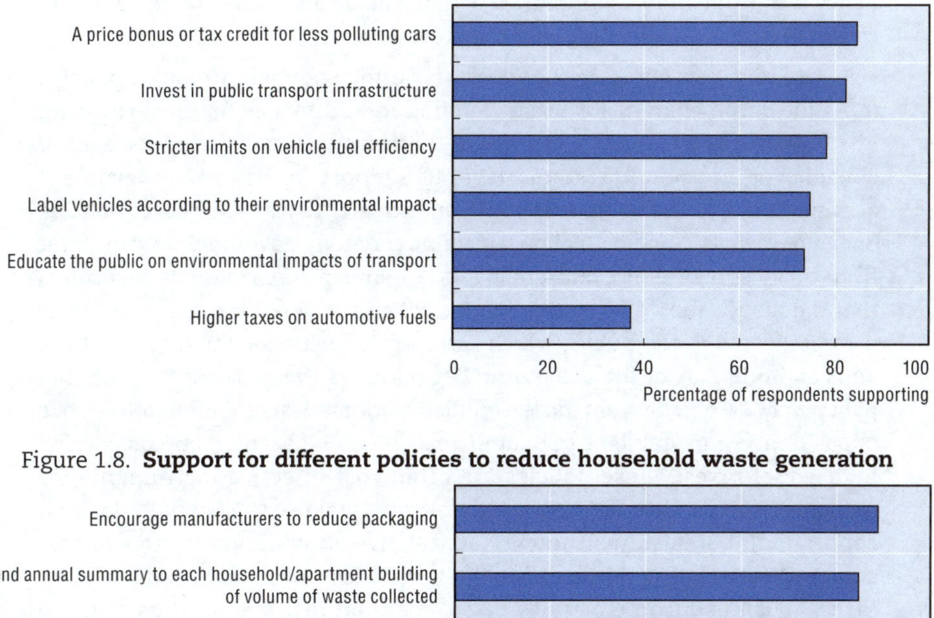

Figure 1.8. **Support for different policies to reduce household waste generation**

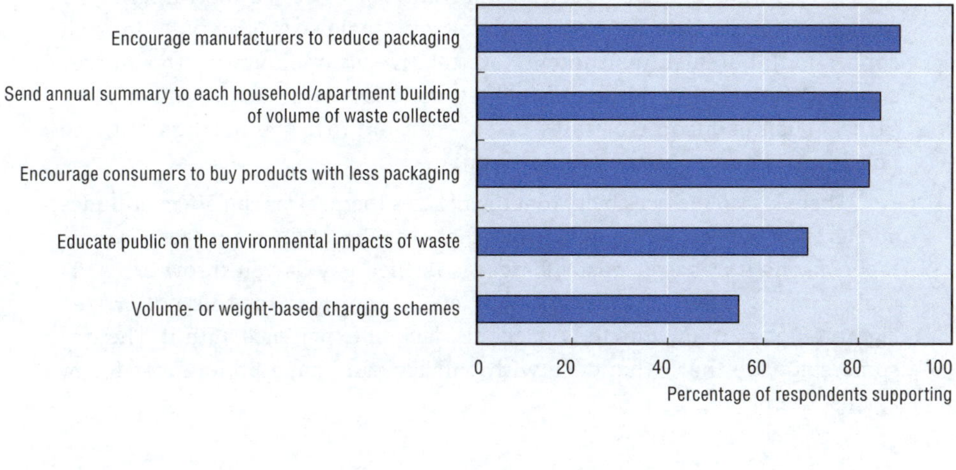

Unsurprisingly, households exhibit the most support for measures in which they stand to gain, and the least support for measures for which they will be liable. This is especially true if these gains and liabilities are pecuniary (subsidies and taxes). Households also exhibit a high level of support for policies which hold other entities liable for environmental externalities. For example, of the waste measures considered in the survey, the one yielding the highest support was persuading manufacturers to use less packaging in their products. Among transport policies, the third most highly supported policy is increasing the stringency of vehicle fuel efficiency standards.

However, it should be noted that, although they are the least popular policy compared to other measures, environmental taxes still garner significant support. Over 35% support higher fuel taxes as a means of lowering emissions and over 50% support unit-based charging schemes for waste. Furthermore, respondents appear quite aware of the potential role of "behavioural" and information-based interventions: The second most popular waste reduction policy is to send annual summaries to households about the volume of waste collected from their dwelling over the previous year.

The fact that some of the most cost-effective environmental policies (e.g. unit-based charges for waste generation and higher fuel taxes) garner less support relative to other policies in the survey stresses the importance of considering households' stated level of support for various government actions. What households and governments identify as the "best" policies does not always coincide. Policies recommended by governments can differ from those supported by households for a number of reasons. One reason is that the justification and motivation for policies may not be fully or properly communicated to the public, which can lead to misunderstanding about the aims and rationale of the policy. Another reason is that, although a proposed policy may be efficient, unaddressed distributional issues can lead to public opposition. An example of this latter reason can be seen when considering higher fuel taxes. These policies are often cost-effective instruments for reducing negative externalities associated with fuel use (e.g. air pollution and congestion), but such instruments are relatively unpopular, as this survey shows. Offsetting the burden of fuel tax increases with reductions in other taxes – such as income or sales taxes – can improve the fairness of these policies, and may enhance popular support.

There is also the possibility that the policies themselves can affect attitudes and levels of support for them. The fact that over 15% of the survey respondents report facing unit-based waste charges (so-called pay-as-you-throw or PAYT) makes one wonder whether the low level of support for this policy (see Figure 1.8) is partially due to households' lack of experience with it. There is some evidence that experience with policies can shape preferences for or against them.

Figure 1.9 shows the percentage of respondents who express some level of support for unit-based waste charges (i.e. PAYT) according to whether they are subject to such charging systems. Experience with PAYT appears to lead to higher support for such systems: In Japan – where PAYT use, at 35%, is second only to Korea (Figure 1.1) – those using such systems are 20% more likely to support the use of these systems than households unfamiliar with them. A similar pattern holds for the other countries in the survey with significant levels of PAYT implementation. The pattern suggests that, although respondents may be resistant to PAYT *ex ante*, it is possible that these attitudes can change once people have more direct experience with such a policy.[28]

The notion that attitudes and preferences towards policies may change after implementation is often not considered in policy analysis, which is that environmental policies targeting individuals facilitate more or less the translation of attitudes into corresponding behaviours. That is, for economists, preferences are normally assumed to be fixed and not subject to additional cues or prompts. As a simple example, consumers with pro-environmental attitudes are predisposed to recycle, but may not have access to services which enable them to translate this attitude into behaviour.

Figure 1.9. **Support for unit-based waste charge policies by exposure to such policies**

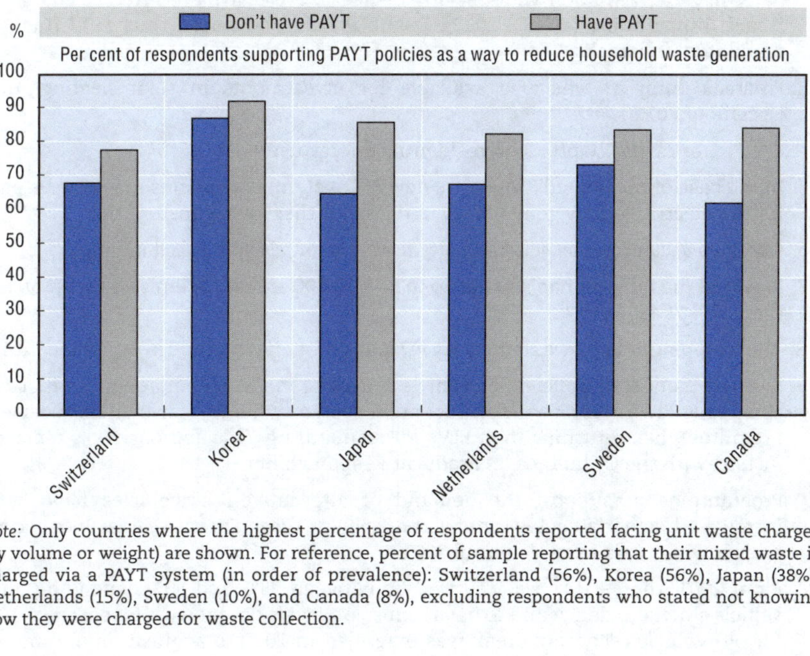

Note: Only countries where the highest percentage of respondents reported facing unit waste charges (by volume or weight) are shown. For reference, percent of sample reporting that their mixed waste is charged via a PAYT system (in order of prevalence): Switzerland (56%), Korea (56%), Japan (38%), Netherlands (15%), Sweden (10%), and Canada (8%), excluding respondents who stated not knowing how they were charged for waste collection.

This view of the interaction between policy and behaviour clearly has some merit, and there are numerous examples where poorly designed policies inhibit pro-environmental actions that households would like to take. Conversely, policy exposure itself may induce a change in attitude – that is, once households are exposed to particular policies, they may be more supportive of them. For example, unexposed households may be more sceptical of unit-based charges for waste collection and water provision than households familiar with such systems. This may reflect households' risk aversion ("The devil you know is better than the angel you don't know").

Chapter 2 focuses more directly on households' norms and attitudes with regard to global and local issues (e.g. economic, social and environmental concerns). The 2011 survey contains numerous measures of these attitudes, which allow the identification of quite complex patterns in views about environmental problems.

Notes

1. While Korean respondents were the most likely to report being charged on a unit basis the most often in the whole sample, the percentage of reported use of PAYT would have been expected to be somewhat greater given the unique waste charging scheme in place in Korea. However, this anomaly may be partly due to the translation and the terminology used to indicate the PAYT option.
2. Early in 2011, 2 million water-saving devices were distributed free of charge to Israeli households (OECD, 2011). It should also be noted that dual flush is mandatory in Israel.
3. Financial support was also available for investments in solar heating until December 2011.
4. See Figure 3.6 in Chapter 3 on residential energy use.
5. In the case of the United Kingdom's new FIT system, households may also be paid for the electricity they generate, even if it is for their own consumption.
6. *http://www.livinggreener.gov.au/rebates-assistance/nt/electricity-feed-in-tariff.*
7. Provided the new car had a value up to EUR 30 000 and a CO_2 emission level of less than 120 g/km.
8. The scrapping scheme was temporary in Japan and came to an end in September 2012.
9. Sweden pioneered policies for vehicle recycling in Europe, enacting its first car scrappage law in 1975 with an incentive system whereby car owners receive a premium when returning their ELVs for dismantling. This approach was replaced in 1997 with the Ordinance on Producer Responsibility.
10. Programmes to support the demand of alternative-fuel vehicles have been introduced as early as 1998 under the *Japanese Clean Energy Vehicles Introduction* project which was extended until 2003.
11. From 2001, the car tax was reduced by between 25% and 50% depending on a vehicle's fuel efficiency and exhaust emission levels, and it was increased by 10% for old vehicles. The tax break was extended in 2009 to acquisition and motor vehicle tonnage taxes.

12. The 2011-2012 *LPG Vehicle Scheme* offers a grant of AUD 2 000 towards the purchase of a new vehicle already fitted with LPG or a grant of AUD 1 250 for the LPG conversion of a new or used vehicle. The grant for LPG conversions is reduced to AUD 1 000 since 2012 for the duration of the scheme to June 2014.
13. In Tessin, for instance.
14. D'Haultfoeuille, Givord and Boutin (2011), "The Environmental Effect of Green Taxation: the Case of the French 'Bonus/Malus'", Direction des Études et Synthèses Économiques, INSEE.
15. A second target of 95g CO_2/km is included for 2020.
16. This mandatory standard will set a national target for average carbon dioxide emissions per kilometre driven and each motor vehicle company will have to contribute to this target.
17. France is also transposing the EU Directive 2010/30/EC on energy efficiency labelling which recasts Directive 92/75/EEC.
18. This approach is to be compared with the absolute comparison method, used for instance in France, where the energy efficiency/CO_2 classes are defined by fixed values.
19. Directive *1999/94/EC* of the European Parliament and of the Council of 13 December 1999 relating to the availability of consumer information on fuel economy and CO_2 emissions in respect of the marketing of new passenger cars.
20. See *http://europa.eu/legislation_summaries/internal_market/single_market_for_goods/motor_vehicles/interactions_industry_policies/l32034_en.htm*
21. The "fuel economy label" indicates the car's fuel economy rank (1 to 5) and its fuel economy (km/L). This labelling scheme replaces the fuel economy rating identification introduced by the government in 2005.
22. According to the packaging law effective as of July 2011 in Israel, producers have to mark each item of packaging with information including the designation of the packaging for recycling, recovery or reuse and the content of hazardous materials in the packaging.
23. The display of the logo on products indicates that fillers, distributors and importers of household products have contributed to the *Eco-Emballages* system, a private company accredited by the French public authorities to install, organise and optimise the sorting and selective collection of household packaging.
24. KRAV standards fulfil the EU standards for organic production in the regulation (EC) No 834/2007 and in some cases are even stricter.
25. Referred to as "green" contracts in Sweden.
26. Electricity labelled with "Bra Miljöval" belongs to this category.
27. Information displayed can range from electricity consumed to current amounts of renewable electricity generated and electricity sold or purchased when applicable.
28. Of course, there are many reasons why people – including environmentalists – could oppose PAYT systems. For example, they could oppose a "commodification" of an environmental issue that crowds out moral imperatives for waste reduction. See, for example, Gneezy, Uri, and Aldo Rustichini. "A Fine is a Price." J. Legal Stud. 29 (2000):1.

Greening Household Behaviour
Overview from the 2011 Survey

Chapter 2

General household attitudes towards the environment

by
Zachary Brown, Nick Johnstone *and* Ysé Serret-Itzicsohn*

This chapter focuses on survey findings concerning households' attitudes towards the environment and how socio-demographic factors such as age, education or relative income relate to environmental attitudes and values. Involvement in volunteer organisations and trust in different information sources are also examined in relation to environmental attitudes. In addition, the importance of environmental concerns relative to other sets of global issues is examined, as well as the perceived importance attributed to different environmental issues such climate change or natural resource depletion. Variations in respondents' satisfaction with different attributes of their local environment (air, water, waste, among others) are also examined. Cluster analysis is applied to the survey data revealing that the respondents can be grouped into three large categories: the environmentally motivated, environmental sceptics and technological optimists.

* OECD Secretariat, Environment Directorate.

The statistical data for Israel are supplied by and under the responsibility of the relevant Israeli authorities. The use of such data by the OECD is without prejudice to the status of the Golan Heights, East Jerusalem and Israeli settlements in the West Bank under the terms of international law.

1. Using general attitudes and beliefs to design environmental policy

The way individuals perceive the importance of different environmental problems relative to other problems and to other priorities for public policy, can strongly affect household behaviours (Millock and Nauges, 2010; OECD, 2011). As the following chapters illustrate, there is wide variation in household choices across the five policy areas analysed in the survey, even when households are subject to the same policy. General attitudes towards the environment and the role of public policy can help explain these varied behavioural responses to policy. For example, people concerned with the environment are more likely to use municipal recycling services (a pattern confirmed in the survey data). The question for policy makers is thus: *Who are these environmentally concerned citizens, and how do they acquire these attitudes?*

This chapter describes general patterns in responses to these attitudinal questions. Identifying how attitudes are linked to socio-demographic characteristics of households can indicate where environmental attitudes cluster in different populations (Olofsson and Öhman, 2006). Such analysis can also generate hypotheses about how these attitudes may change in the future, based on projections of changing socio-demographic patterns (Kahn, 2002). As will be seen below, attitudes and beliefs about the relative importance of environmental issues, and ways of addressing them, vary systematically with respondent age and gender, and in some cases with economic status, household composition, and level of education.

A descriptive analysis suggests that attitudes such as the relative concern and sense of responsibility for environmental problems may evolve in parallel, with ageing populations for example. Furthermore, levels of trust placed in different sources of information are found to correlate closely with environmental attitudes, e.g. concerns about climate change. Environmental quality – especially for global issues such as climate change – has many properties of a credence good, i.e. a good whose value to households depends on third-party assessment. Thus, it is not surprising that the trust placed in different expert sources would determine the perceived seriousness of environmental problems. How this trust evolves in response to the actions of governments, businesses, and the scientific community – and to environmental trends which can be directly observed by households – can affect how popular support for environmental policy accumulates over time, and how it can be lost.

Motivated by these insights, the 2011 round of the Environmental Policy and Individual Behaviour Change (EPIC) Survey included questions on households' perceptions and beliefs on the following subjects:

- The importance of environmental problems relative to other global issues (e.g. social, economic, security-related).
- The relative importance of specific environmental problems, such as natural resource depletion, air pollution, waste generation, climate change, water pollution, as well as endangered species and biodiversity loss.
- The fairness and effectiveness of strategies for solving environmental problems.
- The trustworthiness of different information sources.
- Satisfaction with local environmental quality.
- Knowledge about the causes and consequences of climate change.

Some general behavioural patterns indicative of attitudes were also included in this round of the survey: voting in national and local elections, and participation and contributions to volunteer groups and charities.

In addition, respondents were presented with a set of statements about environmental issues to which they were requested to indicate their degree of agreement or disagreement, for example about their beliefs regarding the potential for technological innovation to solve environmental problems.

In broad terms, much of the data presented below point to a high level of concern for environmental problems relative to other issues, but with economic concerns being uniformly viewed as more important. While views on the relative importance of climate change are closely linked to those on the importance of environmental concerns in general, in some ways climate change is viewed by many respondents as a qualitatively different issue relative to other environmental problems such as resource depletion. For example, the stated level of trust in different sources of information is negatively correlated with stated concern for climate change. Respondent's age also plays a qualitatively different role (see below.)

The data on general environmental attitudes also exhibit qualitatively different patterns across the eleven countries surveyed. For this reason the following analysis presents summary statistics for each country rather than for a pooled sample. This approach identifies country-level differences in policy-relevant relationships between different variables in the survey. For example, in some countries (*as in* Japan) but not others (Spain), respondents' stated favourability towards government environmental policies is highly correlated with whether they vote. However, further work is being undertaken to assess the relative importance of such differences.

2. Perceived importance of environmental concerns relative to other global issues

Early in the questionnaire, respondents were asked to rank six broad policy areas according to what they viewed as "the most serious issues facing the world today" (see Question 22 in Annex A). The policy areas presented were: international tensions, economic concerns, environmental concerns, health concerns, social issues and personal safety.

An important statistic is the frequency with which each of the six policy areas were ranked as the most serious – that is, the percentage of respondents who prioritised one policy area above the others. In every country, "economic concerns" were most frequently ranked as the highest priority out of the six options (Figure 2.1). In seven of the eleven countries surveyed, "environmental concerns" were among the three most frequently ranked as being the most serious. They were less of a priority for respondents in France (fourth most prioritised), Israel (fifth), and the Netherlands (fifth). In Spain they were the least frequently prioritised set of issues. Koreans ranked environmental concerns as the most serious with the greatest frequency (28%). In Sweden, respondents ranked them as the most serious, with the same frequency as economic issues (25%), representing the most even split of opinion among any of the eleven countries.

Concern for the environment relative to the other five sets of issues was clustered among specific socio-economic and demographic cohorts, with significant cross-country differences in the cohorts most frequently ranking

Figure 2.1. **Respondents' three most prioritised world issues, by country**

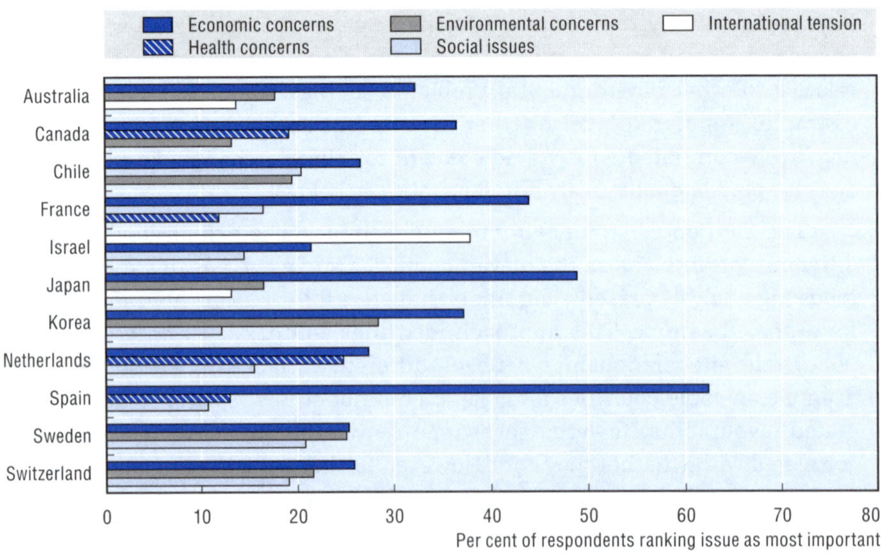

environmental issues as of high importance. In most countries, female respondents were significantly more likely than men to rank environmental concerns among the three most important global issues. Yet in France, Israel, the Netherlands and Spain, this pattern was reversed (Figure 2.2a). In eight countries, younger respondents expressed higher concern for environmental issues, whereas in Israel, Japan and Korea, older respondents were relatively more concerned with these issues (Figure 2.2b).

The empirical relationship between levels of education and concern for the environment is more subtle. In six countries the frequency with which environmental concerns were ranked in the top three issues increased unambiguously with the level of post-secondary education (Figure 2.2c). However, in Israel, Japan and Switzerland, a U-shaped relationship was observed between environmental concern and level of post-secondary

Figure 2.2. **Percentage of respondents ranking environmental issues among the three most serious**

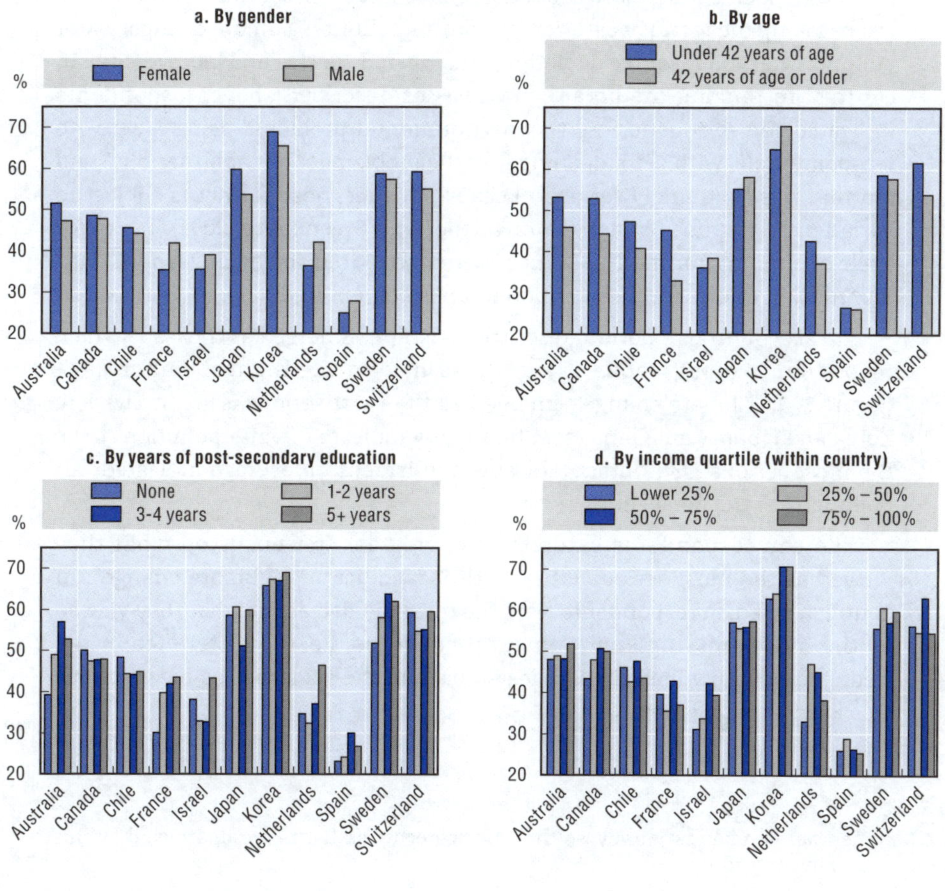

education: Respondents with three or four years of post-secondary education ranked environmental concerns in the top three with less frequency than respondents without post-secondary education, and also those with five or more years of post-secondary education. In Canada and Chile, respondents without any post-secondary education were the most likely to rank environmental issues among their top three concerns.

Similarly, wide variation was shown across countries in the relationship between income and stated concern for the environment (Figure 2.2d). In Australia, Canada, Israel and Korea, respondents with higher incomes were more likely to rank environmental issues among the top three most serious issues. In the Netherlands, concern for environmental issues was highest among middle-income respondents. In the remaining countries, the relationship between income and environmental concern was ambiguous.

3. Perceived seriousness of specific environmental concerns

Respondents were asked about the seriousness of six specific environmental issues facing the world (waste generation, air pollution, climate change, water pollution, natural resource depletion, endangered species and biodiversity). In contrast to ranking the broad global issues described above, respondents indicated the seriousness of the environmental problems using a zero- to ten-point scale, with zero meaning that the problem is "not at all serious" and ten that it is "extremely serious". Because this response format is subject to variation in response styles (Baumgartner and Steenkamp, 2001), ranks for each of the environmental issues were calculated for each respondent.* Summary statistics for these imputed ranks are reported here.

In six countries, "natural resource depletion" emerged as the issue which respondents most frequently cited as having the highest importance (Figure 2.3). "Climate change" emerged as the most serious issue in Sweden, Korea and Japan. Canadians most frequently indicated "water pollution" to be the most serious environmental issue, and Israelis on average indicated "air pollution" as the most serious.

Korean respondents exhibited the greatest consensus on what they viewed as the most serious issue, with 31% indicating "climate change". In contrast, Swedish respondents were the most divided about what they viewed as the most pressing environmental concerns, i.e. the difference in the frequencies with which they indicated each of the six issues as most serious was lowest among them.

* Ties were broken randomly, so that the aggregate statistics are unaffected by this ranking procedure.

Figure 2.3. **Top three environmental concerns indicated to be the most serious**

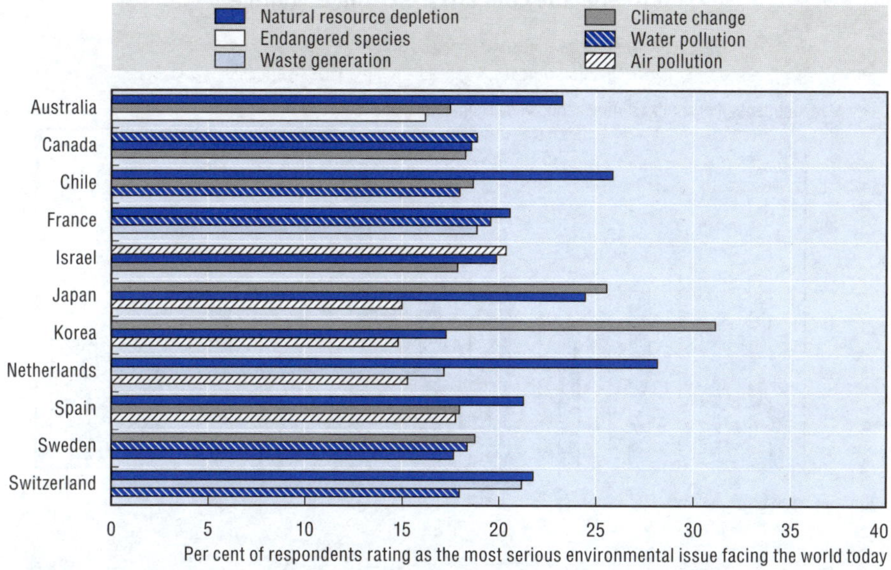

Note: Respondents were asked to rate the importance of six different global environmental issues, each on a 10 -point scale: waste generation, biodiversity loss, air pollution, climate change, water pollution, natural resource depletion, and endangered species. This figure shows the top three issues that were most frequently rated the highest out of the six considered.

Some respondent characteristics were found to be associated with the relative importance attributed to the six environmental issues, in particular climate change. Figure 2.5 shows how frequently younger and older respondents indicated climate change as the most important global environmental issue. Older respondents in all countries except France appear to view it as a relatively more important issue than younger ones. Of course, this means that older respondents rank the other five issues relatively lower than younger respondents. However, no systematic relationship was evident between the relative importance of the other five issues and respondent age. Climate change appears to be a qualitatively different environmental issue for respondents, and one that reflects intergenerational differences in perceptions.

In all the countries surveyed, those who were relatively more concerned with environmental issues relative to other global issues viewed climate change as the most serious of the six environmental issues listed (Figure 2.4). This pattern was most pronounced in Japan, Sweden, and Australia. Initial analysis suggests that, when accounting for overall environmental concern, the relative importance of the six environmental issues is homogeneous across demographic groups.

Figure 2.4. **Relationship between general environmental concerns and specific concern for climate change**

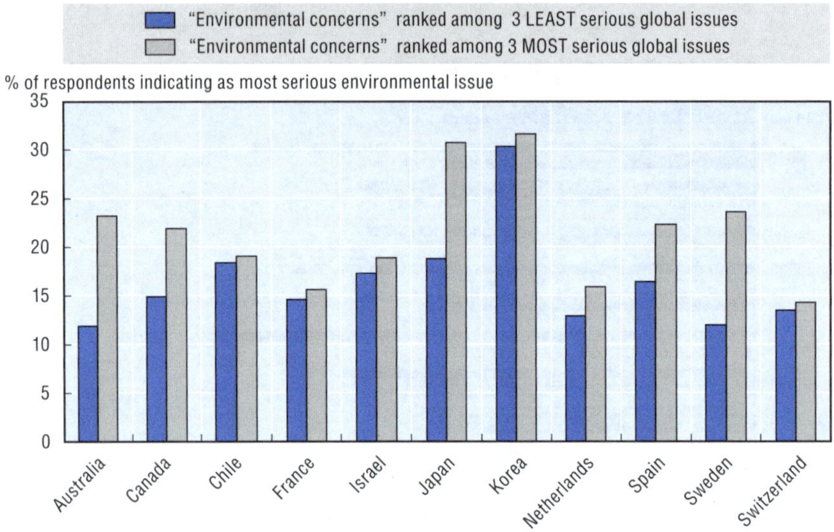

Note: See Figure 2.3 above for a list of all environmental issues respondents considered, and the survey response format.

Figure 2.5. **Respondents reporting "climate change" as the most serious environmental issue facing the world, by age and country**

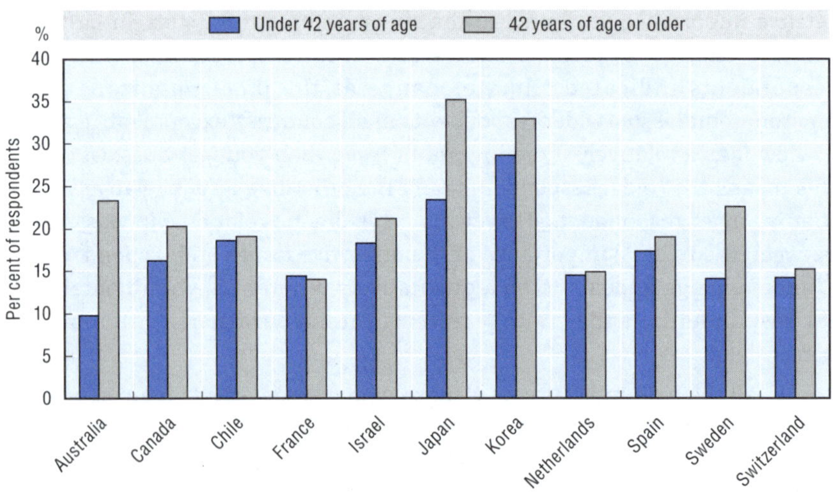

4. General trends in environmental attitudes

Respondents were asked whether they agreed with seven statements addressing different aspects of the environment. These statements, presented in Figure 2.6, cover issues such as reciprocity (i.e. willingness to make sacrifices as long as others do the same), the role of technology in environmental problems, intergenerational equity, and scepticism about claims with regard to environmental issues.

In all but one of the countries, the statement with which respondents agreed the most was: *I am willing to make compromises in my current lifestyle for the benefit of the environment*. Agreement with this statement was highest in Korea, where nearly 95% of respondents expressed a willingness to make such sacrifices. The exception was Japan, where the statement garnering the most agreement was: *Protecting the environment is a means of stimulating economic growth*. In all countries, a majority of respondents agreed with this statement, and also that *government policies to address environmental issues should not cost households extra money*.

The statements with the least agreement exhibit the most cross-country variation. In seven countries, respondents most often disagreed with the proposition that environmental problems were primarily the responsibility of

Figure 2.6. **Levels of agreement with seven statements about environmental policy**

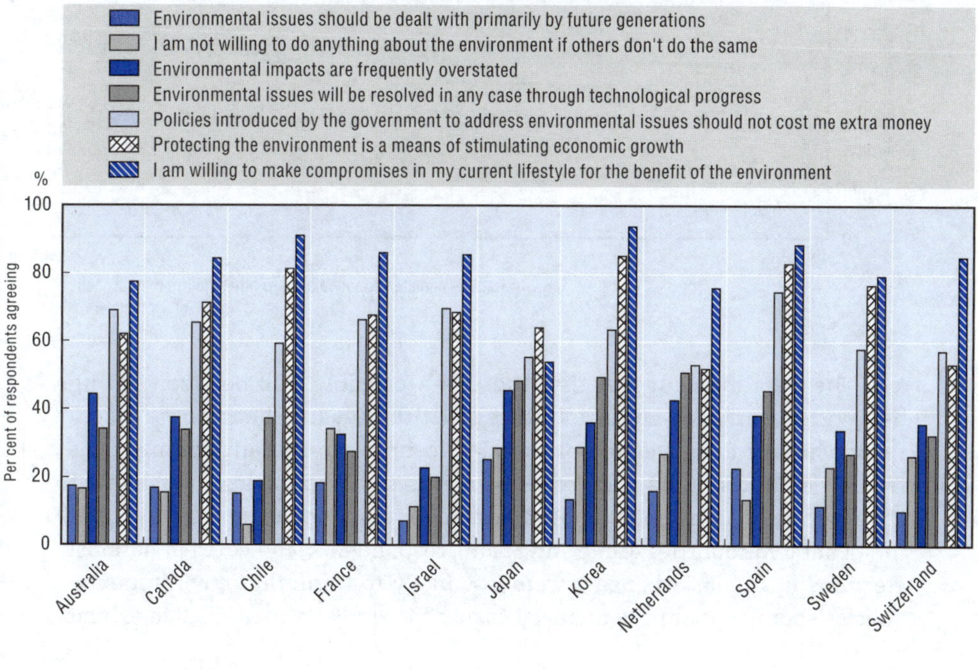

future generations. In the other four countries – Australia, Canada, Chile and Spain – respondents disagreed most with the notion that reciprocity from others was necessary for them to help improve the environment.

The perceived relative seriousness of the environmental issues discussed above also turns out to be closely linked to agreement or disagreement with the seven statements in Figure 2.6. For example, scepticism about environmental claims is closely linked with whether climate change is viewed as a uniquely important problem, relative to the other environmental issues (Figure 2.7). In all countries, respondents who indicated climate change as more important than the other environmental issues were more likely to agree that environmental impacts were frequently overstated. Australians' views on climate change were most closely tied with scepticism about environmental claims, whereas Chileans were the least so.

Figure 2.7. **Scepticism about environmental impacts and views on the seriousness of climate change**

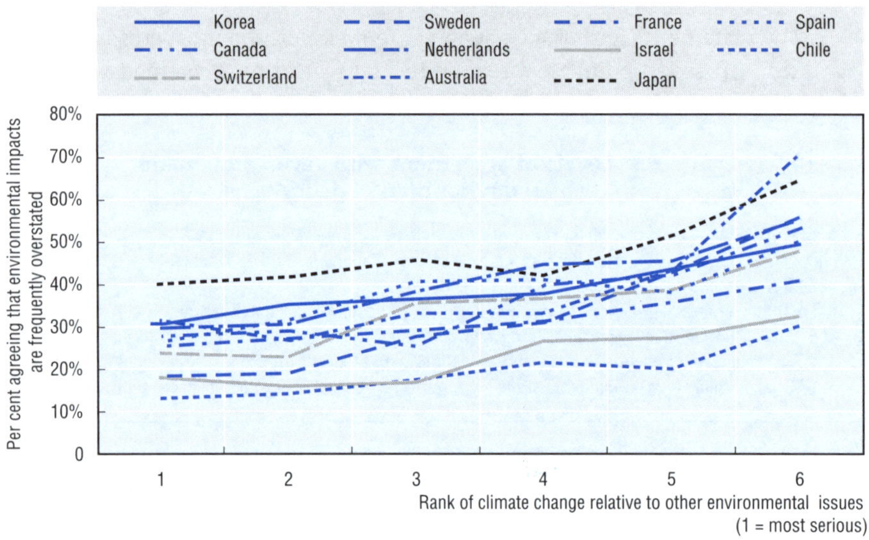

The data also suggest that there is a possible link between voting behaviour and respondents' stated willingness to bear some fiscal responsibility for government policies to improve the environment. As Figure 2.6 shows, a majority of respondents in all countries thought that environmental policy should not cost them extra money. It can be noted that for all but two countries (Chile and Spain), respondents who voted in national elections in the last six years were less likely to think that environmental policies should not impose financial costs on them. The Japanese data exhibit the largest difference between voters and non-voters in this regard. Further econometric analysis is necessary to confirm the robustness of this apparent link between voting behaviour and willingness to accept a financial burden to fund environmental policies.

It is striking to note that in six of the eleven countries, concerns about intergenerational equity appear to be greater among older respondents (Figure 2.8). That is, older respondents more frequently expressed a belief that their own generation bore significant responsibility for solving environmental problems – i.e. that such problems should not be simply left for future generations.

Figure 2.8. **Views on intergenerational equity across ages**

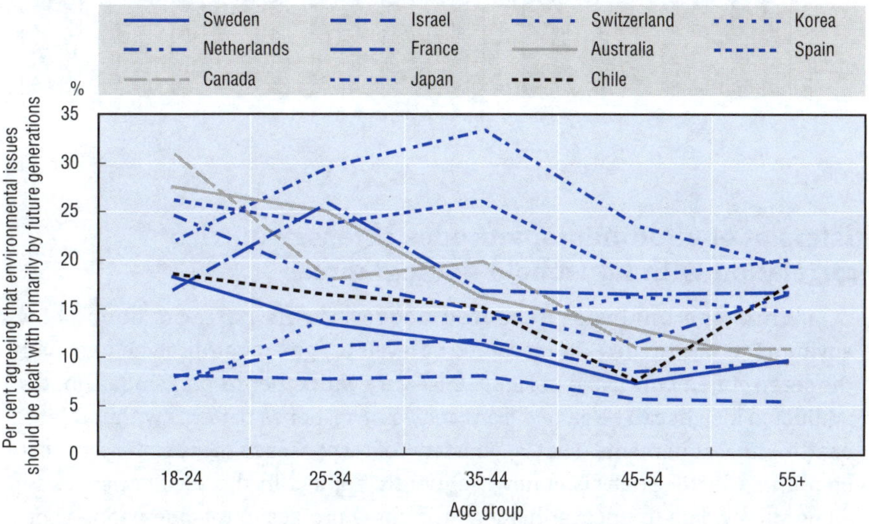

Respondents who participated in or supported volunteer organisations or charities expressed less of a need for reciprocal action on the part of others in return for their own effort at reducing environmental impacts (Figure 2.9). In all countries except Spain, respondents who declared participating in or supporting environmental organisations expressed the greatest willingness to take unreciprocated action for improving the environment. These relationships are most prominent among French respondents for whom expressed preferences for reciprocity were the highest. As for Swiss respondents, their rates of support/participation in environmental organisations were the highest (20%), while the rates of Japanese respondents were the lowest (3%).

Figure 2.9. **Agreement with the statement "I am not willing to do anything about the environment unless others do the same", by participation in volunteer organisations**

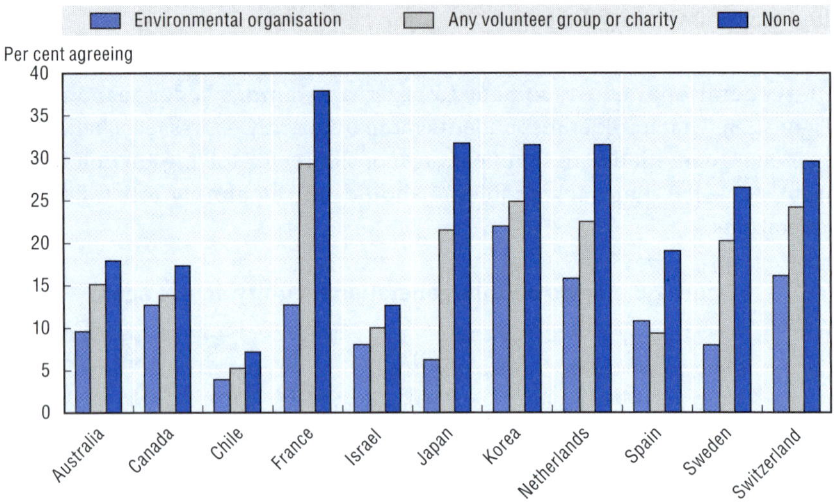

5. Clusters of environmental attitudes across countries and correlation with household demographics

While the analysis presented above reveals patterns about how environmental attitudes – as measured by levels of agreement/disagreement with the seven statements listed in Figure 2.6 – vary with other respondent attributes, additional insight can be gained from analysing response patterns within this set of attitudinal statements. That is, "clusters" of responses to these statements can be found. Cluster analysis is now frequently applied in the social sciences for using survey data to uncover hidden attitudinal profiles in the population about different issues. For example, environmental attitudes in a given population can be basically arranged into two groups, with environmentalists on the one hand and environmental sceptics on the other – indeed, it shows how agreement/ disagreement with the seven attitudinal statements could reflect this binary pattern. However, it is also possible that those attitudes towards the environment are more complex, and that this complexity is reflected in the set of responses to the seven attitudinal statements.

To uncover these attitudinal profiles in an objective manner – without introducing bias from the analyst – a statistical method known as latent class analysis (LCA) is used. A description and demonstration of LCA in the context of environmental attitudes can be found in Morey, Thatcher et al. (2006). LCA produces a number of useful statistical results: First, it can provide a statistically sound indication of how many attitudinal profiles – henceforth referred to as "classes" – can best represent the data at hand; secondly, it provides an estimate of likely responses to each of the seven attitudinal statements for an average member in each class; and finally, LCA provides an indication of which class each respondent most likely belongs to.

A preliminary application of LCA to the 2011 survey data suggests that there are three attitudinal classes which generate the responses to the seven statements in Figure 2.6. Table 2.1 presents the results from the LCA. These classes have been labelled *environmentally motivated*, *environmental sceptics*, and *technological optimists*. Note that the statistical procedure also revealed a fourth class corresponding to "extreme responders," who disagreed (or indicated "no opinion") with every one of the seven attitudinal statements. Respondents in this class were estimated to embody 2% of the sample, and tended to complete the survey much faster than other respondents. This class is therefore set aside for the remainder of the analysis in this section.

The *environmentally motivated* comprise just under half of the pooled sample, and make up the largest of the four classes. They believe that environmental problems are real and express a willingness to make compromises in their lifestyle to solve them. Members of this class also expressed the least need for reciprocation from others in order to undertake action to solve environmental problems. *Environmental sceptics* believe that environmental issues are overstated and do not wish to pay for government environmental policies. But, on the other hand, they do report a general willingness to make compromises for the benefit of the environment, though not to the same degree as the other two substantive

Table 2.1. **Percentage of respondents in agreement with seven attitudinal statements, by class membership**

%

		Classes			
	Pooled sample	*Environmentally motivated*	*Environmental sceptics*	*Technological optimists*	*Extreme responders*
Class size (percent of sample)	*100*	*46*	*32*	*20*	*2*
Policies introduced by the government to address environmental issues should not cost me extra money	63	57	77	61	0
I am willing to make compromises in my current lifestyle for the benefit of the environment	84	92	77	90	0
Protecting the environment is a means of stimulating economic growth	71	74	69	80	0
Environmental issues will be resolved in any case through technological progress	38	0	58	100	0
Environmental impacts are frequently overstated	35	0	100	0	0
I am not willing to do anything about the environment if others don't do the same	21	11	38	17	0
Environmental issues should be dealt with primarily by future generations	16	8	27	16	0

classes. *Technological optimists* share the belief with the *environmentally motivated* cluster that environmental problems are real and appear willing to make lifestyles compromises to solve them. The key difference between the two clusters is that the first group expresses a greater belief in the potential of technological progress to solve environmental problems.

LCA provides not only the summary statistics presented in Table 2.1, but also indicators for which class each respondent belongs to. This permits analysis of how the size of each of these attitudinal classes varies across countries (Figure 2.10) and by demographic characteristics (Figure 2.11). The environmentally motivated are most common in Israel. Technological optimists are most prevalent in Chile and Korea, and the least so in Sweden. In Japan, the Netherlands, Spain and Australia, environmental sceptics comprise the largest of the three attitudinal classes. The belief most strongly linked to the sceptic class is that environmental claims were overstated (see Table 2.1). Furthermore, respondents could be classified as environmental sceptics and still be relatively willing to bear some financial burden to fund government environmental policy (as in the Netherlands and Japan, when comparing Figures 2.6 and 2.10); it would seem that if some sceptics believed in the alleged impacts of environmental problems, they would more likely support the costs of policies to abate them. Turning to demographics, gender turns out to be a significant determinant of class membership, with women uniformly more likely to be classified as environmentally motivated than men. Age also is associated with environmental attitudes, but the direction of the trend varies across countries (Figure 2.11b). In Australia, France, Korea and Switzerland, membership in the sceptic class increases with age, while in Japan this trend is reversed. Chile exhibits a clear U-shaped trend, with the

Figure 2.10. **Size of environmental attitude clusters, by country**

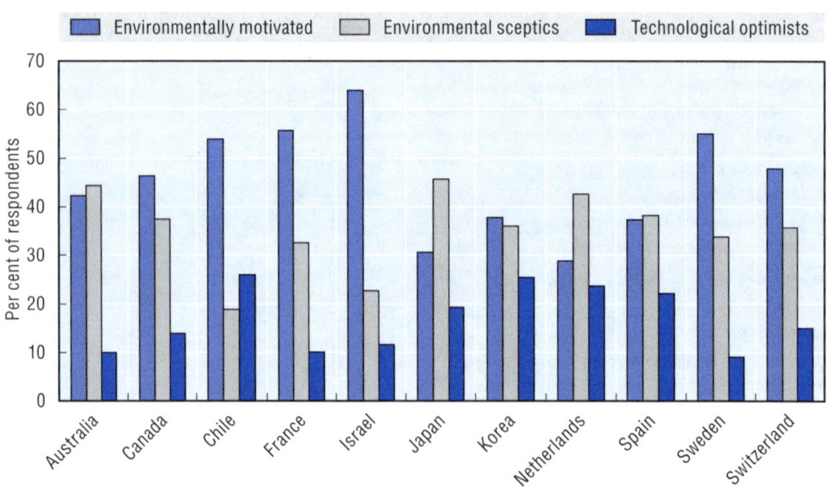

frequency of sceptics first decreasing with age until their 30's, and then increasing. The other countries exhibit no clear trend between these variables.

It is important to remember that these clusters represent a classification of environmental *attitudes*, not *behaviours*. Although this chapter also focuses exclusively on attitudinal data, it is a particularly important point here, because there is a danger of assuming that intentions map directly onto actions, and therefore that the size of the *environmentally motivated* class – 46% of the pooled sample – appears implausibly large. Indeed, the statements corresponding to membership in this class are often at odds with the demonstrable prevalence of much environmentally harmful behaviour (e.g. a stated willingness to make lifestyle sacrifices and take unreciprocated action to improve environmental conditions, while at the same time exhibiting inefficient use of energy and water, throwing significant amounts of unused food away, etc.). But in fact membership in the *environmentally motivated* class reflects only how people wish to see themselves, how they wish others to see them, or how they intend to act. What this discussion – in combination with subsequent chapters – indicates is that this likely mismatch between intentions and actions is policy-relevant, because it suggests a role for policies which eliminate barriers to households who intend to "follow through" on environmentally beneficial actions.

Figure 2.11. **Size of environmental attitude clusters, by respondent characteristics**

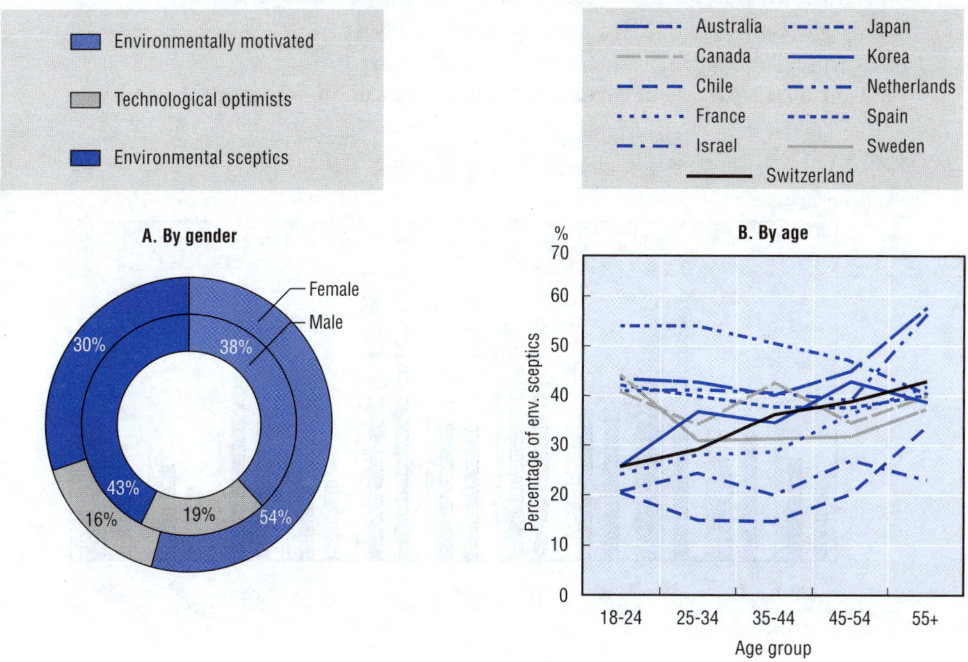

6. Respondents' satisfaction/dissatisfaction with aspects of their local environment

Respondents were also asked how satisfied they were with their local air and water quality, their access to "green spaces" such as parks and forests, the level of noise in their neighbourhood, and the level of litter and rubbish in their area. The response format consisted of a 5-point Likert scale, ranging from "very satisfied" to "very unsatisfied", and including a "no opinion" option. Attention is focused here on when and why respondents were at all unsatisfied with specific aspects of the local environmental quality.

Overall, Koreans, Israelis and Chileans most frequently expressed dissatisfaction with their local environment (Figure 2.12). In Korea, a majority of respondents were dissatisfied with four of the five local environmental aspects addressed. The Dutch in general appear the most satisfied. In comparing levels of dissatisfaction across countries, it is important to recognise that many factors, aside from the environmental aspects themselves, can affect whether or not an individual is satisfied. These factors can include one's general outlook on life and cultural norms regarding the expression of unhappiness or dissatisfaction (Marris, Langford et al., 1998). Somewhat surprisingly, litter and rubbish problems have the highest percentage of dissatisfied respondents in five of the countries surveyed – the most frequent of any of the local environmental aspects covered. Conversely, in eight countries, access to green space is the aspect with the lowest levels of dissatisfaction.

Figure 2.12. **Levels of dissatisfaction with local environmental quality**

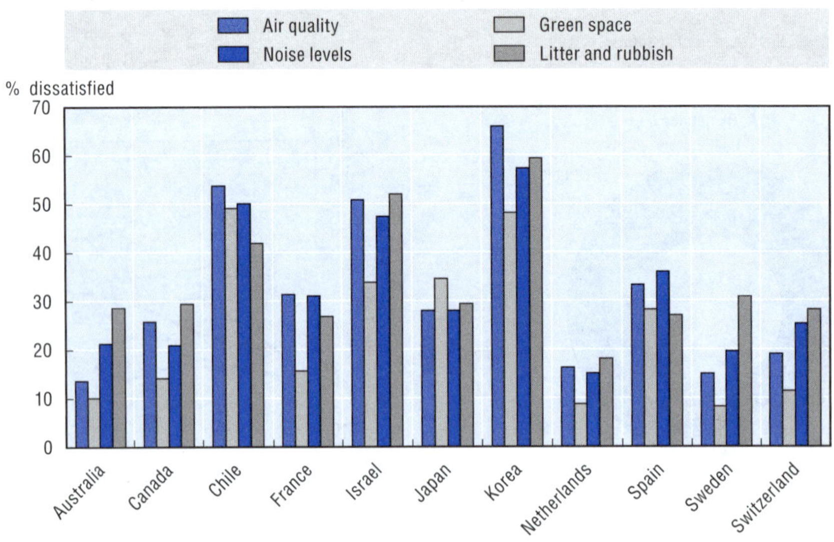

Intuitive patterns can be seen in the relationship between respondents' stated satisfaction with different attributes of environmental quality and, for example, population density. Figure 2.13 demonstrates this for local air quality: Across all countries, respondents in urban or suburban areas more often expressed dissatisfaction with their local air quality than their counterparts in rural areas. This difference is greatest in Korea, which also happens to have the highest overall level of dissatisfaction with local air quality. A similar though somewhat less pronounced trend (not shown) holds for the other four local environmental aspects covered in the questionnaire.

Figure 2.13. **Dissatisfaction with air quality and household location**

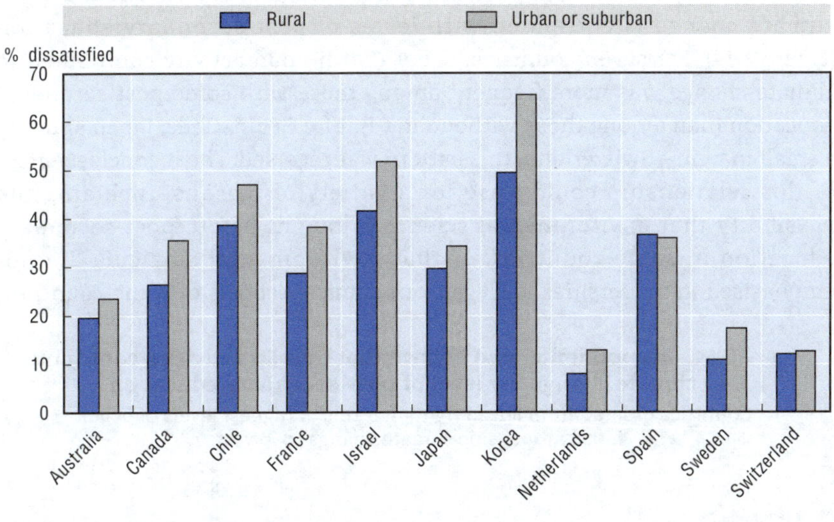

7. Knowledge and beliefs about climate change

Additional questions were introduced in the 2011 round of the survey to elicit respondents' knowledge and beliefs about the causes of climate change (see Q29 in Annex A). In particular, respondents were asked to indicate whether or not they thought the following statement was true:

Every time we use coal, oil or gas, we contribute to climate change.

In every country, a clear majority of respondents believe that climate change is at least partly caused by human activity. Somewhat strikingly, the Dutch respondents were most sceptical, with only 64% agreeing with the above statement. Koreans were the most likely (90%) to believe it. However, this finding is less surprising upon re-examination of Figures 2.1 and 2.3: Within the Dutch sub-sample, "environmental concerns" are not in the three most often cited global issues ranked as most serious (Figure 2.1), nor was "climate change" most frequently indicated to be the most important environmental issue (Figure 2.3). In contrast, both environmental concerns in general and climate change in particular emerge as relatively serious issues for the Korean sub-sample.

One would expect that respondents' belief in the above statement should be influenced by their knowledge of environmental issues, since scientific evidence establishes that climate change is indeed anthropogenic. Furthermore, we would expect that environmental knowledge increases with respondents' levels of formal schooling. However, one would also expect that respondents' beliefs depend on their level of trust in different information sources – whether they believe the scientific information they receive.

Surprisingly, there is mixed evidence in the data that belief in anthropogenic climate change is correlated with levels of post-secondary education (Figure 2.14). For seven countries, belief that human activity contributed to climate change was more frequent among those with some post-secondary education than among those without any. But for Spain, Chile, Japan and – by a small margin – Switzerland, this pattern was reversed. The inconclusiveness of this relationship could arise for a variety of reasons, including the possibility that environmental science is taught before post-secondary education in some countries, or that environmental curricula are not emphasised in universities, colleges, or technical schools of some countries,

Figure 2.14. **Respondents who believe that human activities contribute to climate change, by level of post-secondary education**

Countries ordered from left to right, in terms of highest levels of belief in anthropogenic climate change to lowest.

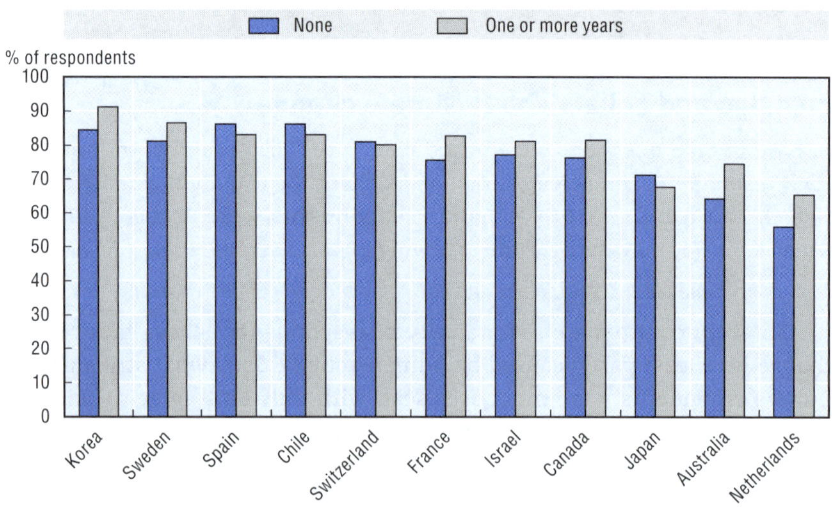

or else that respondents in some countries do not trust the scientists and organisations claiming that climate change is at least partly caused by anthropogenic factors.

The survey included questions on respondents' level of trust in information from *i)* scientific experts, *ii)* government, *iii)* environmental non-governmental organisations (NGOs), *iv)* consumer organisations, and *v)* manufacturer organisations. Specifically, respondents were asked to indicate, using a 10-point scale, how much they trusted claims from these sources regarding the environmental impacts of products.

Figure 2.15 shows that the level of trust respondents place in scientific experts is strongly linked to whether they believe that climate change is caused by human activity. This pattern is strong in ten of the countries surveyed, and in the eleventh country – Korea – the direction of the correlation is the same, but very weak. A similar association was found with respect to trust in government, environmental NGOs, and consumer organisations. No significant relationship was found between trust in manufacturers/retailers and beliefs regarding anthropogenic climate change. These statistics suggest that disbelief in the anthropogenic aspects of climate is due less to low exposure to scientific information about climate change, and more to overall trust in received information. This preliminary finding highlights an important area for further analysis.

Figure 2.15. **Respondents who believe that human activities contribute to climate change, by level of trust in scientific experts**

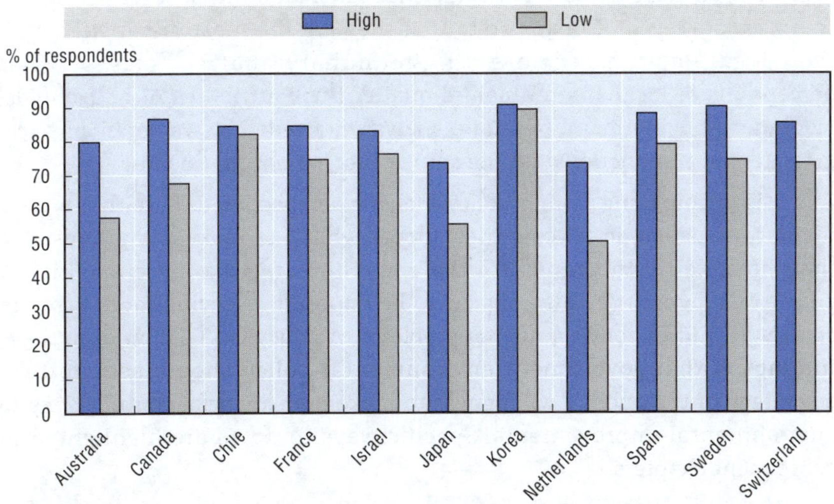

Note: "High trust" is calculated as being equal or above the within-country median on the 10-point scale for that question; "low trust" is below the within-country median.

8. Conclusions

The results presented above clearly demonstrate that environmental issues are viewed by a majority of the surveyed populations as a relatively high priority relative to other global concerns. In most of the countries, women viewed environmental concerns as more pressing than men did, but exceptions include France, Israel, the Netherlands and Spain. Younger respondents tended to view environmental problems as more pressing, but again exceptions were found for Canada, Israel, Korea and the Netherlands. On the other hand, neither relative income nor respondents' level of post-secondary education seems to serve as a significant predictor of environmental concern.

Furthermore, concerns about climate change clearly stand out as qualitatively distinct from other environmental issues. Older respondents and those who view environmental issues as relatively important are evidently more likely to view climate change as the most pressing environmental issue. And whether or not people view climate change as one of the more important environmental problems in the world is closely related to the scepticism with which they view environmental claims, and the trust they place in scientific experts.

At the local level, there appears to be significant variation across countries as to the most problematic environmental aspects of households' communities. Litter and rubbish emerges as the dominant concern in five of the countries, but each of the covered aspects of local environment – with the exception of noise levels – ranks as least satisfactory in at least one of the countries surveyed. Koreans state that they are the most dissatisfied with their local environment overall. Preliminary analysis suggests that perceptions of local environmental quality track attributes of households' biophysical environment, but more analysis is needed by way of biophysical indicators in conjunction with the survey data to confirm this finding.

It is encouraging from a policy maker's perspective to note that a clear majority of people in all countries appear willing to make compromises in their lifestyle for the benefit of the environment. However, a majority of people in all countries also appear to be opposed to paying more taxes or charges to address environmental problems. A policy implication from these findings is that government environmental policy should seek to take advantage of the public's willingness to make non-financial contributions to environmental improvements. Specific ways to do so are highlighted in subsequent chapters.

Aggregate analysis of the attitudinal data from the survey indicates that respondents can be grouped into three basic "clusters" of views towards the environment: Just under half of the sample exhibits a high level of motivation to improve the environment, unilaterally if necessary. They state a relatively high willingness to make compromises in their lifestyle to solve environmental problems. Over a third of the sample, on the other hand, displays sceptical attitudes towards environmental problems, expressing doubt as to whether such problems actually exist. The remainder of the sample view environmental problems as real but solvable principally through technological innovation. Of most interest with respect to these general clusters of attitudes is that the size of each cluster varies dramatically across the surveyed countries. For instance, environmental sceptics comprise over 45% of the Japanese sample, but less than 20% of the Chilean sample. The 2011 EPIC Survey has thus captured a wide variety of environmental attitudes across the countries surveyed. The reasons for these cross-country differences in attitudes point to a policy-relevant topic for future analysis.

References

Baumgartner, H. and J.-B. E. M. Steenkamp (2001), "Response Styles in Marketing Research: A Cross-National Investigation", *Journal of Marketing Research,* Vol. 38, No. 2, pp. 143-156.

Kahn, M. E. (2002), "Demographic change and the demand for environmental regulation", *Journal of Policy Analysis and Management*, Vol. 21, No. 1, pp. 45-62.

Marris, C. et al. (1998), "A Quantitative Test of the Cultural Theory of Risk Perceptions: Comparison with the Psychometric Paradigm", *Risk Analysis*, Vol. 18, No. 5, pp. 635-647.

Millock, K. and C. Nauges (2010), "Household Adoption of Water-Efficient Equipment: he Role of Socio-Economic Factors, Environmental Attitudes and Policy", *Environment and Resource Economics*, Vol. 46, pp. 539-565.

Morey, E., et al. (2006), "Using Angler Characteristics and Attitudinal Data to Identify Environmental Preference Classes: A Latent-Class Model", *Environmental and Resource Economics*, Vol. 34, No. 1, pp. 91-115.

OECD (2011), Greening Household Behaviour: The Role of Public Policy, OECD Publishing. doi: *10.1787/9789264096875-en.*

Olofsson, A. and S. Öhman (2006), "General Beliefs and Environmental Concern", *Environment and Behavior*, Vol. 38, No. 6, pp. 768-790.

Greening Household Behaviour
Overview from the 2011 Survey

Chapter 3

Household behaviour and energy use

by
Bengt Kriström*

This chapter reviews the evidence collected in the survey on households' energy-related behaviour and their responses to various types of policies targeting renewable energy and energy efficiency. It also examines differences in behaviour across households and the effect of norms and attitudes, such as the perception of environmental issues.

* Umeå-SLU University, Sweden.

The statistical data for Israel are supplied by and under the responsibility of the relevant Israeli authorities. The use of such data by the OECD is without prejudice to the status of the Golan Heights, East Jerusalem and Israeli settlements in the West Bank under the terms of international law.

1. Introduction

This chapter reviews the evidence collected in the survey on households' energy-related behaviour and in particular their responses to various types of policies targeting renewable energy and energy efficiency. The policy measures reviewed range from economic incentives, such as energy conservation grants for home insulation or financial support to install solar panels, to the provision of information to consumers with the labelling of appliances' energy efficiency. In addition, governments (and service providers) can differentiate the characteristics of energy-related services through electricity tariffs which vary according to the time of the day, or the installation of smart electricity meters giving households access to real-time information so as to guide their consumption choices. The chapter also examines differences in behaviour towards energy use across households' economic and demographic characteristics (income, age, education). And finally, the effect of norms and attitudes, such as the perception of environmental concerns, is analysed as well.[1]

This chapter provides a better understanding of the main determinants of households' behaviour towards residential energy use in order to improve the design of public policies. Drawing upon observations from over 12 000 respondents in 11 OECD countries, this descriptive analysis of the responses to the EPIC survey provides a first picture on ways to promote energy efficiency and an increased use of renewable energy sources at home.

Some of the key questions addressed in this chapter include:

- How do general attitudes towards the environment (environmental awareness; membership in environmental organisation) correlate with demand for energy efficiency and for renewable energy?
- Who invests in energy efficiency measures? Who takes advantage of grants or subsidies for investments in energy conservation?
- Who invests in renewable energy equipment (e.g. solar panels)? Who takes advantage of grants to install/use renewable energy?
- What is the impact of energy efficiency labelling for appliances and, where available, for buildings?
- How much are households willing to pay to use only renewable energy? Does WTP vary significantly across household groups?

The following section sheds light on electricity spending patterns and explores the link between income and electricity expenditures. Then are examined households' energy use in their residence and choices of different options offered by electricity providers, such as differentiated and green tariffs. The other sections examine the demand for renewable energy, including respondents' willingness to pay (WTP) for renewable energy; the factors driving energy-saving at home, the use of government support schemes and the role of labelling. A summary of key findings on energy-related choices that households make at home is provided in the box below.

Box 3.1. **Residential energy use: key findings**

Findings from descriptive analysis suggest that:

- There is significant, unmet household demand for electricity generated from renewable sources: 60% of respondents are willing to pay extra for their electricity to be generated from renewable sources, but 45% who express an interest in having differentiated rates for renewable energy do not have this option.
- Household demand for electricity decreases with the price of electricity and increases with household income (as expected), but both effects are relatively inelastic.
- This means that, without additional policy measures, higher energy prices will have disproportionate welfare impacts on low-income households, in line with other studies. In addition, complementary policies may be required to support price-based policies.
- Economic and attitudinal factors simultaneously determine energy-saving behaviour: higher-income households on average engage less frequently in energy-saving behaviours, but respondents living in households who frequently engage in energy-saving behaviours have a higher concern for environmental problems relative to other global issues.
- Governments in the surveyed countries play an important role in promoting household investments in energy efficiency, but there is wide room for additional interventions: 16% of the energy efficiency investments recorded in the survey received government financial support. Energy efficiency labels also play a role in conservation: across the countries surveyed, households who recognised energy efficiency labels for appliances spent on average 6% less on electricity than households who did not recognise these labels.

2. Households' electricity consumption and spending patterns

Reported expenditures on electricity

The residential and commercial/public sectors together consume about 60% of electricity output in OECD countries,[2] following a steady growth path from its 46.5% share in 1973.[3] The OECD residential sector consumes about as much electricity as the commercial/public sector and has roughly tripled its consumption from 1973, while consumption of the industrial sector has dropped (IEA/OECD, 2010). The electricity share in total energy consumption varies substantially between countries, including in the residential sector where its share has increased. Electricity prices for residential consumers also vary across OECD countries and are typically higher than for industrial consumers. The average household price for electricity was USD 0.20 per kilowatt-hour in 2009 with strong variations across the OECD (from USD 0.77 per kWh in Korea to USD 0.365 per kWh in Denmark).[4]

In general, electricity bills contain a fixed and a variable part, and various types of subsidies and rebates exist. For example, in Australia, the residential customer price list details off-peak load prices, block prices, four different types of green energy tariffs, and eleven different rebates for medical reasons (as for home dialysis).[5] Chile grants subsidies for electricity consumption for lower-income households and Sweden varies the energy tax geographically with four different pricing zones since 2011. In summary, the cost of each additional kWh consumed depends on many factors that vary across countries.

The literature on energy demand at the household level notes the existence of significant variations in consumption, driven by differing household characteristics.[6] Price and income remain the key drivers of spending patterns. This was confirmed by an analysis of the previous round of the EPIC Survey (OECD, 2011). However, other factors (suchas household composition, dwelling characteristics, and others) also play a role. Cultural differences and social norms have also been shown to underlie part of the inter-country differences in energy use patterns.[7]

The survey responses allow us to examine reported residential energy demand and the average price households pay for electricity. Respondents were asked how much approximately was their total annual electricity consumption in their primary residence during the last year. It should be noted that energy demand is a derived demand in the sense that it is used to run appliances, for lighting, space heating and cooling, and for water heating. In particular, energy is usually combined with a capital good (such as a washing machine) to provide a service, and this has a number of implications. First, technological progress means that the amount of energy needed for a given use can be reduced.

However, this does not necessarily result in reduced energy consumption and could even increase it as per the Jevons paradox.[8] As a matter of fact, energy efficiency improvements lower the effective price of energy, which increases the quantity demanded. Secondly, because the relevant capital stock is typically fixed in the short run, the household's response to an energy price change is likely to be significantly smaller than the long-run response. Thus, it takes time before a policy has an effect because it takes time for households to adjust their real capital stock. In short, the fact that energy demand is a derived demand has important policy implications as well as important consequences for the economic analysis of household response.[9]

According to the survey data, the reported mean electricity spending is about EUR 962 per year and the average budget share is about 3.5%. This is consistent with what is known from household expenditure surveys in OECD countries. Figure 3.1 provides additional information about the distribution of electricity spending around the mean: For each country, 50% of the data lie within the shaded boxes, and 90% of the data lie within the dotted lines.

Figure 3.1. **Distribution of households' reported electricity spending, by country**

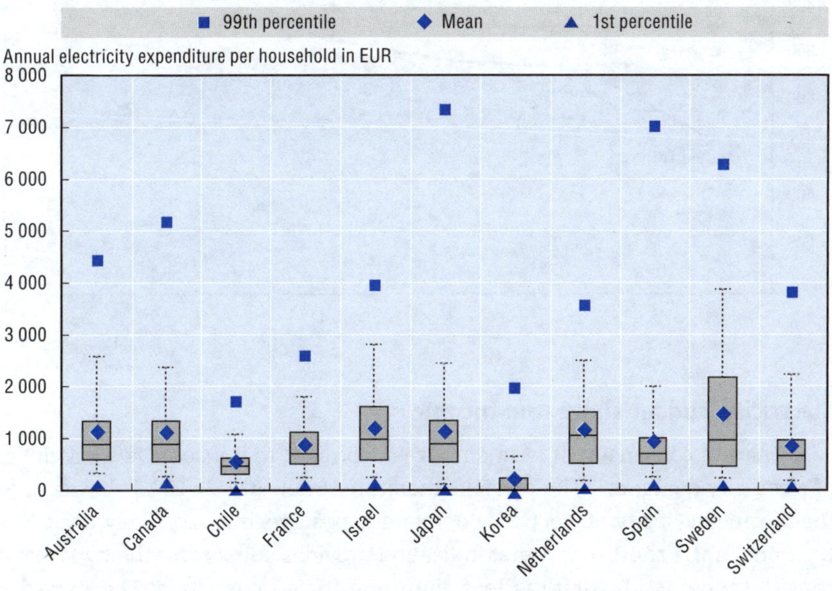

Note: After data cleaning, 6 719 answers could be used to calculate these results, representing a little more than half of the total sample. Lines in boxes represent, from bottom to top, the 25th, 50th (median) and 75th percentiles. The dotted vertical lines (i.e. the "whiskers") in the figure represent 5th (bottom) and 95th (top) percentiles.

Disaggregating the data further, it is found that the mean, the median and the standard deviation of spending are higher in houses than in apartments. This is important since the distribution of housing type varies quite a bit across the countries surveyed. For example, while almost 80% of the Australian respondents live in a detached or semi-detached house, about 70% of Spanish respondents live in an apartment.

Households living in an apartment spend, on average, EUR 700 per year on electricity (or about EUR 60 per month) and those living in a detached or semi-detached house, EUR 300 to 500 more per year (or 25% to 40% more per month). Chilean and Korean households appear to spend significantly less than their Australian counterparts (used as a baseline), while households in Israel, the Netherlands and Sweden spend significantly more on electricity. There is also a strong income effect, such that spending increases with household income. As expected, electricity expenditures increase with household size, but less than proportionately (see Figure 3.2).

Figure 3.2. **Reported annual electricity expenditures, by household size**

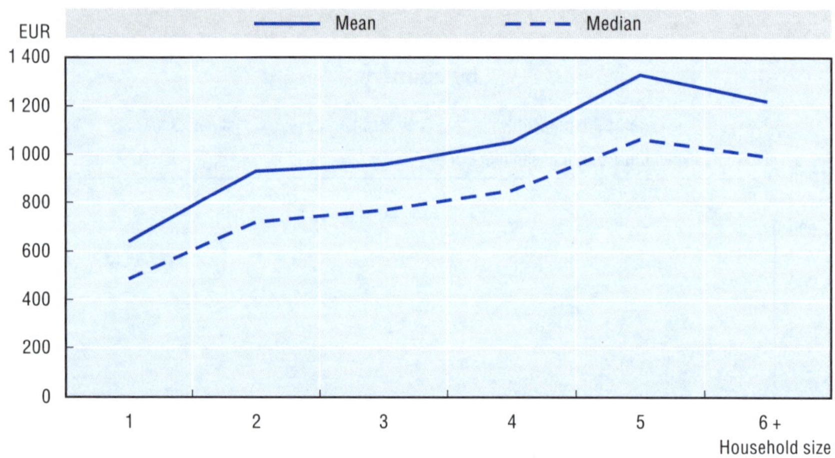

Electricity budget share and income

There is consensus, at least among economists, that income is a key driver of electricity demand.[10] The previous literature review (OECD, 2008) showed that the income elasticity of electricity demand is positive, but very likely to be less than one in the short run. A first indication provides support for the hypothesis that the income elasticity is less than one for all countries. The negative relationship between the electricity budget share and income is clearly visible in Figure 3.3. However, there is some evidence of non-linearities, indicating that income elasticities vary across income groups. Overall, these findings add to the bulk of evidence showing that higher energy prices have regressive effects.

Figure 3.3. **Relationship between household electricity budget share and income, by country**

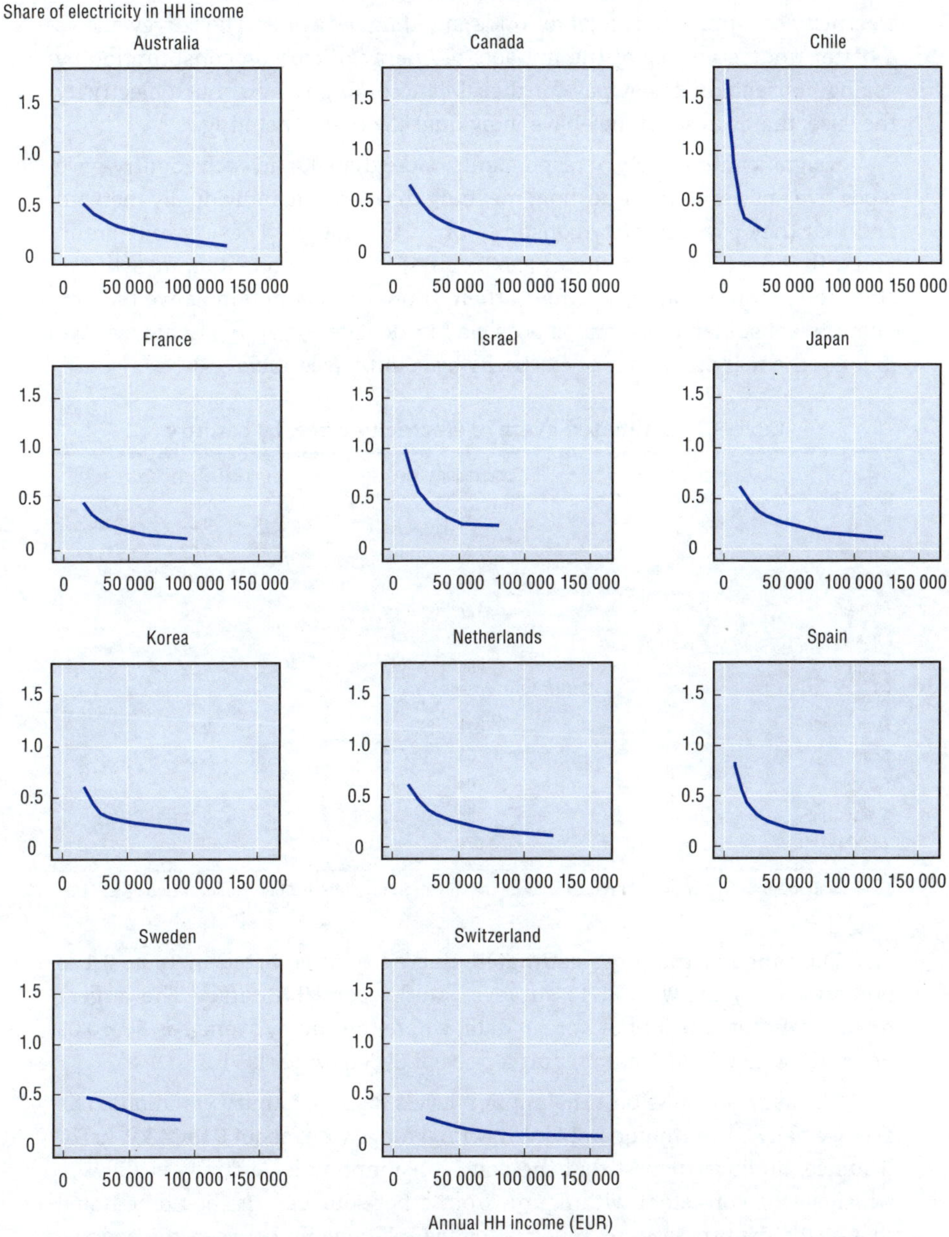

Note: The data are displayed as the predicted values from Lowess-smoothed estimation. Lowess-smoothed estimation is a way to estimate the relationship between two variables which makes fewer assumptions than linear regression statistical techniques. Specifically, Lowess-smoothed estimation finds the best-fitting curve to model the relationship between two variables. This curve can take a variety of shapes, in contrast to linear regression which assumes the best-fitting curve is a straight line.

Average electricity price and consumption

Whether or not households report paying for the marginal unit of electricity consumed is crucial for understanding behaviour. The survey allows a better understanding of the impacts of energy pricing on consumption by asking households if they pay for their bill according to how much electricity they use, that is to say if they have individual electricity metering.

A significant majority of respondents, more than 90% in each country, gave a positive answer to this question, meaning that they are individually metered and that they pay electricity consumption at the margin. These results are in line with those of the 2008 survey (OECD, 2011).[11] As respondents were asked to state the amount corresponding to their annual consumption as well as the quantity consumed in kWh, it is possible to calculate the approximate average (but not the marginal) price of electricity by country (see Table 3.1).

Table 3.1. **Estimated average electricity price, by country**

	Observations	Price (EUR per kWh)
Australia	131	0.22
Canada	133	0.13
Chile	189	0.28
France	177	0.16
Israel	142	0.16
Japan	208	0.22
Korea	70	0.05
Netherlands	241	0.30
Spain	136	0.28
Sweden	294	0.17
Switzerland	158	0.16
OECD (11)	*1 879*	*0.20*

Note: Respondents were not required to answer the question on their annual consumption.

The range of the gross estimated electricity price listed in Table 3.1 is broadly consistent with EU15 surveys[12] and recent IEA data.[13] The largest relative difference is for the Korean data, where the survey average is EUR 0.05 per kWh, while the IEA data report a price of EUR 0.07 per kWh.

The survey results on consumption levels display country variations. The average daily consumption is 17 kWh, with a range from about 9 to 34 kWh. For instance, findings suggest that the daily consumption in Australia is 15 kWh, which seems consistent with figures from other sources.[14] It should be noted that at the country level there is a negative relationship between the average electricity price and the quantity of electricity consumed in kWh as Figure 3.4 shows. Korea is an outlier with its low price and relatively low consumption.

Figure 3.4. **Relationship between average electricity price and electricity consumption**

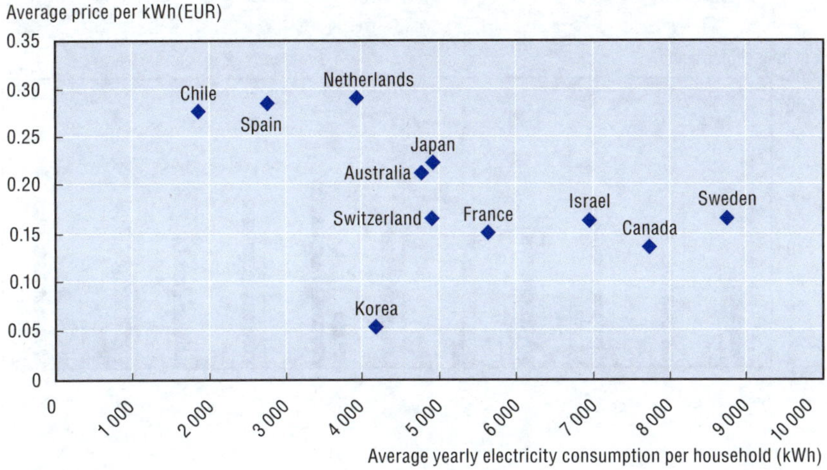

3. Households' energy choices in their residence

Residential energy use typically includes space heating and cooling, water heating, cooking, lighting and the use of appliances and equipment. Other choices depend on what options the energy service provider offers, such as differentiated electricity rates and smart metering.

Choice of energy source

A rough overview of energy sources used by households for space heating and cooling in the eleven countries surveyed is provided in Figure 3.5. District heating, which can occasionally deliver cooling services, appears to be almost exclusively used in Korea, and particularly in Sweden where it represents up to 30% of the energy sources for space heating. Not surprisingly, electricity is the most widespread source for cooling and heating in all countries, with slightly lower figures in Korea, Sweden and Switzerland. If respondents in Israel essentially only use electricity; the picture is a bit more complex in the other countries. For example, while French respondents typically use either electricity or gas and other fossil fuels for cooling and heating, 40% of Japanese respondents claim to use both sources of energy for this purpose.

Turning to the reported energy sources for water heating, the quite extensive use of thermal solar panels in Israel can be noted. Finally, as regards cooking, gas and other fossil fuels dominate in Chile and Korea with over 90% using these fuels, while electricity is the dominant means of cooking in Canada, Sweden and Switzerland. Overall, some 16% of those who responded use both sources for cooking; the combination is most common in Israel.

Figure 3.5. **Reported energy sources for space heating and cooling, by country**

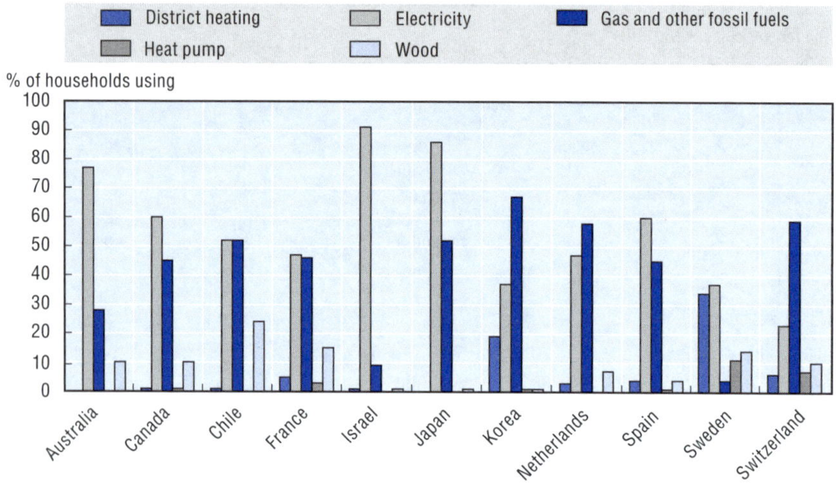

Note: Respondents could tick more than one alternative.

Respondents were invited to indicate whether they themselves had invested in solar panels for electricity generation or for heating water. Such investments are often not feasible for those living in apartments and so the data presented focus on those living in detached or semi-detached houses. As can be seen in Figure 3.6, between 2% and 10% of households possess these technologies in their homes, except in Australia and Israel where investment is much higher. The seemingly large levels of investment in solar panels in Israel is clearly associated with regulations which have been in place since 1976 requiring all buildings up to nine stories high to be equipped with solar water heaters. As expected, installation of solar panels is generally higher among owner-occupied homes, but the differences are not large. However, the figure includes those who have made the investments themselves or those whose residence was already equipped with solar panels when they took up residence. The difference is more striking when the latter are removed from the sample.

A total of 158 respondents (1.3%) indicate that they have invested in wind turbines, with another 126 (1%) stating that their residence was already equipped with a turbine. Ownership rates are highest in Australia (3.3%), Canada (3.1%), Korea (3.3%) and Israel (2.8%). Sweden has by far the highest stated ownership rates of ground-source heat pumps (13.3%).

Services proposed by electricity providers

In order to better understand households' demand for energy-related services which have the objective (at least partly) of reducing the impacts of residential energy use on the environment, respondents were asked which of

Figure 3.6. **Reported installation of solar panels**

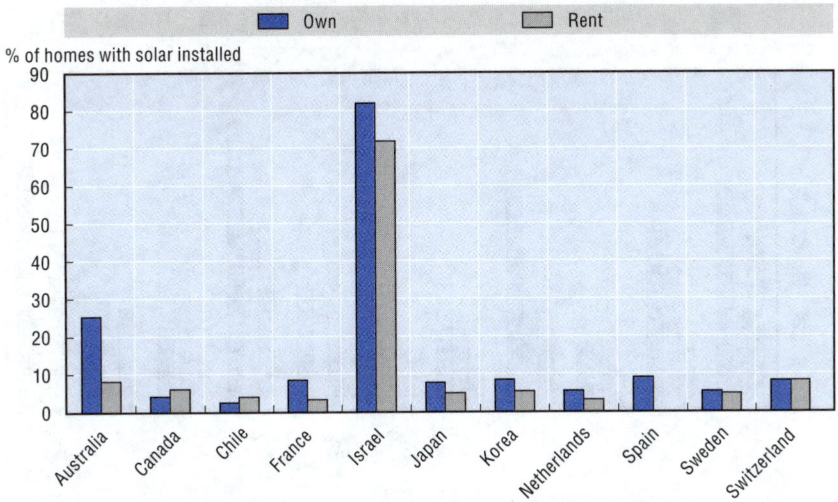

Note: This percentage is calculated for those living in houses, and who report that they themselves have invested or that the house was already equipped.

the following service options had been proposed by their electricity provider (see Question 68 in Annex A): *i)* "renewable/green" energy tariffs where a specific amount of renewable electricity is guaranteed in the supply; *ii)* smart electricity meters allowing them to monitor consumption by viewing electricity usage in real time; and *iii)* reduced tariffs for off-peak hours.

Overall it is in Australia, the Netherlands and Switzerland where such options are provided most frequently (see Figure 3.7). Service provision is greatest for off-peak tariffs in all countries, except Sweden.[15] Green tariffs are most common in Switzerland, the Netherlands and Australia. In the case of smart metering, Canada stands out, followed by Australia. However, it must be noted that some respondents clearly misinterpreted such questions. For example, in Chile none of the three services are available, but approximately 10% of respondents indicated it was available. Conversely, in Sweden "green" contracts are almost universally available although only 40% know that this is the case.

From a policy perspective, it is important to have an idea where latent demand is potentially greatest. Looking more closely at the data, the responses indicate that unmet demand – reported by those who would be interested in the service option but have not been offered the option by their service provider – is greatest for "smart metering". This is particularly true in Spain, Israel and Chile (see Figure 3.8).

With respect to "green" tariffs, the greatest unmet demand is in Korea, Israel, Spain and (particularly) Chile. For off-peak tariffs, the greatest unmet demand is in Sweden, Spain, Korea, Israel and Chile (see Figure 3.9).

Figure 3.7. **Reported provision of special electricity services**

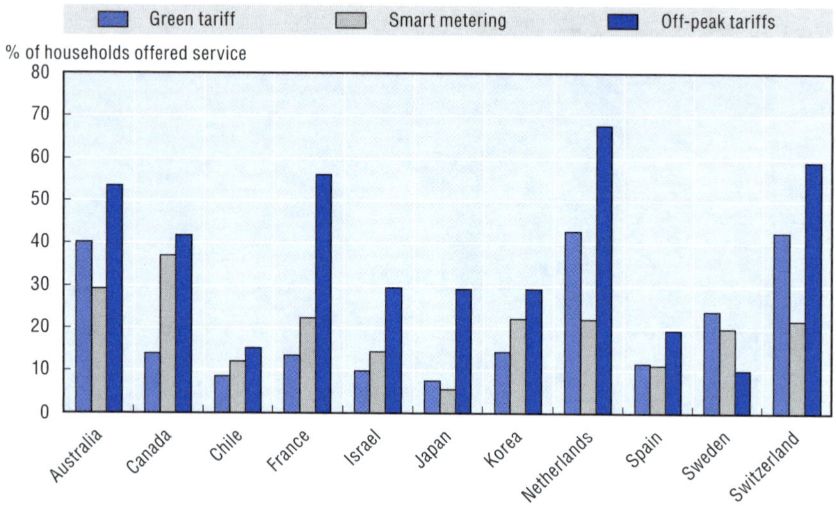

Figure 3.8. **Reported access to smart metering and interest expressed, by country**

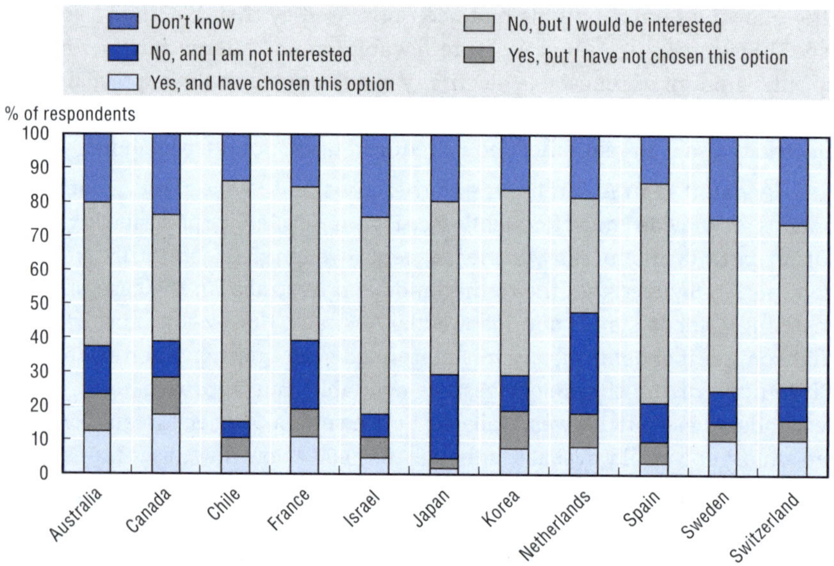

More generally, information failures are also important from a policy perspective. In many cases respondents may not have been aware of the services provided, even if they do have some demand for the service. Five countries – Australia, Canada, Israel, Sweden and Switzerland – have relatively high rates of respondents indicating that they "don't know" if any of the services are provided.

Figure 3.9. **Reported demand for differentiated electricity rates, by country**

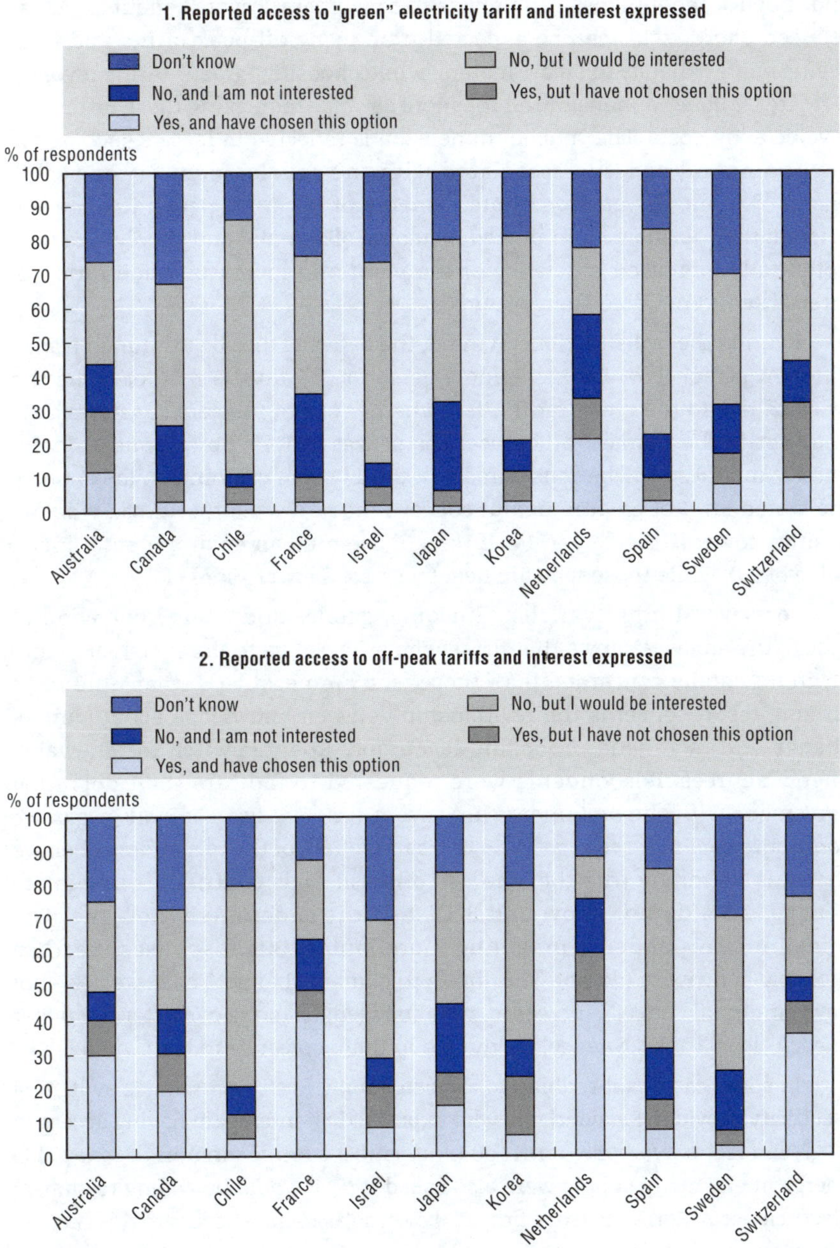

These services differ slightly with respect to the combination of "private" and "public" benefits they can provide to consumers. Private benefits, such as reduced energy bills, can be expected from using off-peak tariffs and from saving energy thanks to smart meters, while choosing "green" tariffs involves a greater "public" dimension, all the more so since requesting these tariffs can involve extra costs. The "public" dimension is reflected in factors such as the environmental benefits associated with the development of renewable energy.[16] While the environmental benefits of lower off-peak rates are not evident, they do allow for improved matching of demand and supply, and thus potentially increased efficiency of the plant and increased penetration of intermittent renewable energy sources.

Bearing these distinctions in mind, the data can also be examined from a different perspective. Let us define a person to be "in-the-market" if he/she was positive to the green tariff offer (whether or not the service is available) and "not-in-the-market" otherwise. This leaves those who responded "don't know", and these will be considered as a third type of customer. A look at how the stated level of environmental concern index reveals that those who are positive towards the "green tariff offer" have a mean environmental index value of 8.0, while those who are negative have a lower score of 7.2.

For several other variables (income, gender and energy behavioural index), the analysis shows that those who are positive to the switch are more environmentally concerned than those who are negative. Perhaps the most striking result concerns the relationship between knowledge about climate change and likelihood of looking favourably upon a switch to renewable energy sources. Respondents were requested to indicate their degree of agreement with the statement that "every time we use coal, oil or gas we contribute to climate change".[17] In this sample 79% of the respondents believe that it is definitely true that the use of fossil fuels contributes to climate change, 7.6% do not know and 3.4% think it is definitely not true. The remaining respondents providing more ambiguous responses, i.e. they probably contribute (do not contribute) to climate change. The stated level of environmental concern increases with knowledge, and is consistently higher at each "level" of knowledge, for the group that is positive to the switch.

In some cases, subscribing to a green supply tariff necessitates switching electricity provider. Respondents who expressed an interest in having access to a "green electricity supply tariff" guaranteeing a specific amount of renewable energy in electricity supply were also asked if they would be willing to change electricity provider, at no extra financial cost, so as to have access to this service. One may wonder why anyone would resist such a switch, assuming that the offered alternatives provide higher environmental benefits; however, the findings show significant differences between the countries, in terms of the proportion of households that will make the "free" switch (see Figure 3.10).

Figure 3.10. **Percentage of households reporting that they would make a "free" switch of electricity provider to benefit from "green tariffs", by country**

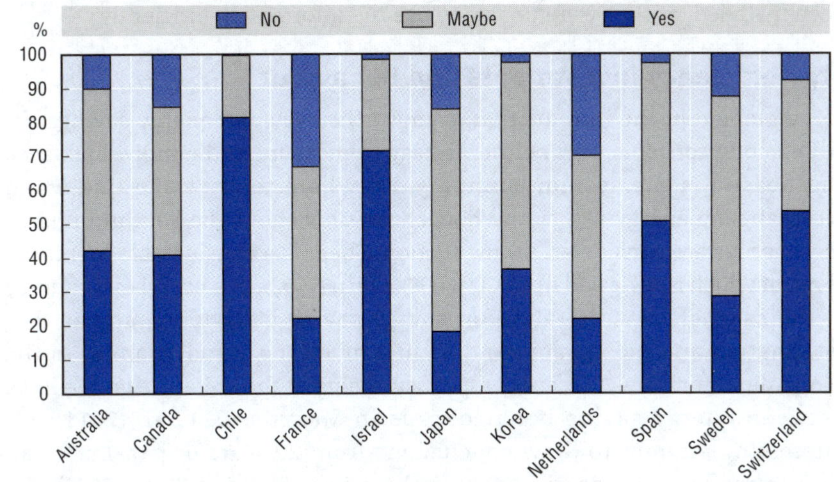

Note: The question was filtered in the sense that it was only asked as a follow-up question to those respondents who expressed an interest in having access to a "green electricity supply tariff" guaranteeing a specific amount or renewable energy in electricity supply. This represents about 46% of the total sample.

While the majority of respondents in Chile, Israel, Korea, Spain and Switzerland indicate that they would definitely be willing to switch, the French and Dutch respondents report that they are less likely to do so with more than 30% of the persons surveyed not willing to change provider. The Japanese respondents appear the most undecided about this potential service. These findings illustrate the existence of barriers (perceived or real) to switching provider, and apparent bias towards the "status quo" even when switching is costless. Attitudinal variables describing environmental concern and propensity to save energy are the most important drivers for making the switch.

A further policy-relevant concern is whether smart metering appears to have any effect on consumption. Over 1 000 respondents report that they have smart meters. Of those, 41% claim that the smart meter has helped reduce their consumption, 24% use such meters but find no effect on their consumption, while 28% say that they do not use it (the remaining 8% provided assorted comments on why they were not using it, even if it was installed). The paucity of the data makes it difficult to come to any hard conclusions on the real effect. There are only 91 valid observations on electricity consumption for respondents who have used smart meters and state that they found it reduced their consumption.[18] Finally, the respondents were also given the option to provide open-ended comments on smart meters. Because there are only 80 different comments, they can hardly give an exhaustive picture of how smart meters are received. The respondents typically stated that the smart meters were about to be installed or had just been installed, or they were unable to use the information efficiently.

4. Energy efficiency investments and behaviour

As noted in the introduction, energy efficiency and energy savings are quite subtle concepts, and neither is easy to measure at the individual level. One reason is the system boundary, i.e. where to draw the line when disentangling energy use by the household. This should be kept in mind as these issues are considered in this section. There is a considerable literature, some of which is reviewed in OECD (2008). Stern (1992), in a useful early survey of the psychological literature on energy-saving behaviours, argues that information and money are handled in a much too casual manner in the economics literature. A recent compact summary of the literature on the links between energy-saving behaviours is in Martinsson et al. (2011). An interesting attempt to draw conclusions from 20 years of pan-European surveys regarding residential energy behaviour is found in Stead (2009), the main point being that energy-saving behaviour appears to be relatively stable.

Some links that have been disentangled in the literature are:

- *Investment vs. non-investment:* Households tend to employ non-investment types of energy efficiency measures, according to Gustavsson et al. (2010). On the other hand, households consistently appear to require a very substantial rate of return in order to invest money in energy efficiency improvements. Why this is so is somewhat of a paradox. See OECD (2008) for a brief review.
- *Age and home-ownership:* Barr et al. (2005) show how households tend to save more and this tendency is stronger among older households. Several studies show that ownership has a positive influence on energy saving (Martinsson et al., 2011).
- *Household income:* Lower income induces relatively more saving (Hedberg and Holmberg, 2005)
- *Environmental concern:* Stronger environmental concern is positively correlated with energy saving (Brandon and Lewis, 1999; Abrahamse and Steg, 2009).
- *Gender:* There is some evidence that women are more likely to undertake measures to save energy (Carlsson-Kanyama and Lindén, 2007).
- *Norms and habits:* Some literature highlights the role of habits – i.e. "not fully conscious forms of behaviour" – in energy saving (Marechal, 2010).

Energy-saving behaviour

A 2011 study by Expertise et Conseils en Maitrise de l'Energie (ECME) interviewed consumers in the EU27 countries about their energy consumption. One question was "what have you done to reduce your electricity consumption at home in the last 12 months?".[19] The most common activity to save electricity is by switching off lights (92% across the full EU sample) and appliances (82%). The ECME report concludes that "consumers are actually quite serious about reducing energy consumption."

The OECD survey asked a slightly different question (see Q76 in Annex A): "How often do you perform the following in your daily life?" The activities listed are:

- Turn off lights when leaving a room.
- Cut down on heating/air conditioning to limit energy consumption.
- Only run full loads when using washing machines or dishwashers.
- Wash clothes using cold water (e.g. 30°C).
- Turn off stand-by mode of appliances.
- Air dry laundry rather than using clothes dryers.

In order to visualise the data in an aggregate form, an index of energy-saving behaviour was created. Figure 3.11 plots the mean scores for this index by country (note that it is only the relative differences in scores across countries that are meaningful in this figure). The scores of Canada, Japan, the Netherlands, Switzerland and, at the bottom, Sweden, are lower than the average. Spain tops the scale. It is tempting to suggest that Spanish

Figure 3.11. **Energy-saving behaviour index by country**

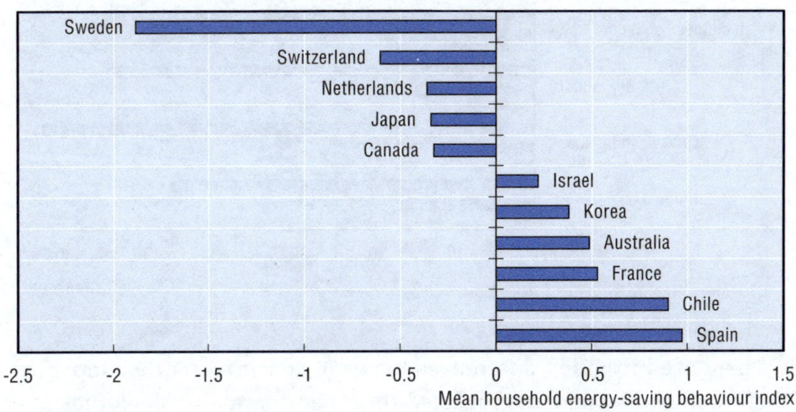

Note: The energy-saving behaviour index used to generate this figure was created as follows: A score was generated for each of the six energy-conservation behaviours, with 0 being the score associated with "never", 1 with "occasionally", 2 with "often", and 3 with "always". The scores for each of these behaviours were then added up into an aggregate score, which was rescaled to have mean zero and standard deviation equal to one.

households are the most likely to engage in energy-saving behaviour. From an economic perspective, this result likely reflects the fact that electricity has been relatively expensive in Spain and the energy-saving behaviour is a consequence of this. The only significant deviation from the theories depicted above is that the index is lowest for the youngest and oldest respondents. It is interesting to note that it drops linearly with higher income, is slightly higher for females and for home-owners; lastly, it increases with the environmental concern index. Furthermore, the data show that as the index increases monotonically, the more willing the person is to change his/her lifestyle for the benefit of the environment (see Q26 in Annex A).

Looking at the individual activities in more detail yields interesting insights (see Figure 3.12). As one moves down the y-axis from "always" to "never", the number of respondents decreases at each alternative. Overall, the general pattern is that respondents' most popular answer to these questions on energy saving is "always", but with some country variations. In particular, Swedish respondents stand out in their energy-saving behaviour as discussed in more detail below. Still, most respondents report that they engage in at least some "energy-saving behaviour", a finding documented in many surveys (e.g. Stead, 2009).

Figure 3.12. **Energy-saving behaviour by practice**

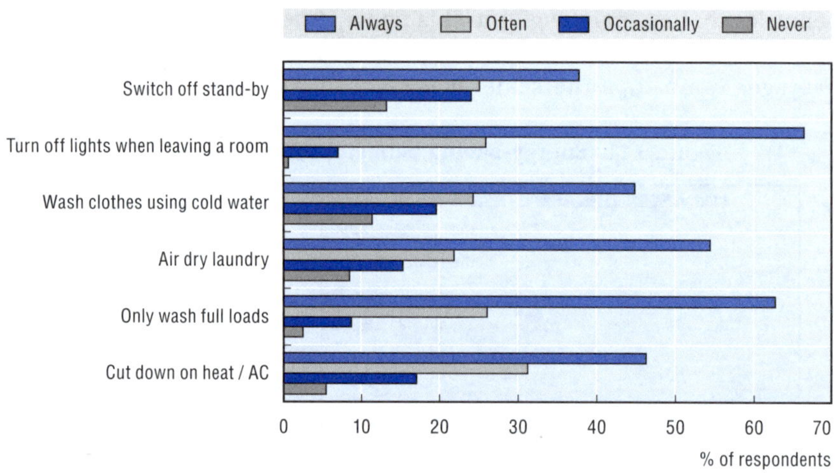

There are intriguing differences between countries. For example, while 45% of Swedish respondents reported that they "never" wash clothes in cold water (less than 30 degrees Celsius), almost 80% of Chilean respondents "always" use cold water. As with many other of the activities listed, the motivation for doing so may be "mixed", and it is not possible to disentangle private (e.g. financial) from public (e.g. environmental) motives. Furthermore, as discussed earlier in this chapter, variation in social norms and cultural factors can explain much of the cross-country differences in these behaviours.

There is interesting information in the survey regarding the use of stand-by devices;[20] 50% to 70% of respondents from EU member states in this survey report that they "often" or "always" turn off the stand-by mode (the Dutch and Spanish respondents being the most active). To sum up: a simple analysis of the data shows that energy-saving behaviour is related to age, income, environmental concerns and willingness to make compromises in people's current lifestyle much in the way as the literature suggests.

Choice of residence and energy conservation

A majority of the respondents claim not to have taken energy costs into account when purchasing or renting their current primary residence, suggesting that there is a potential to bring about change by increasing awareness of energy conservation possibilities for people when they are changing residence. As Figure 3.13 shows, Canadians and Swedes appear to be the most concerned, with roughly 40% reporting that they take these costs into account, compared to one out of four on average for all eleven countries surveyed.[21] Korea comes third (34%) while Spain ranks last with only 13% positive replies. It should be noted that countries where respondents report that they take energy costs into account more often have all introduced energy labels for buildings. In addition, inspection of the data indicates that living in

Figure 3.13. **Percentage of respondents reporting taking into account energy costs when changing residence**

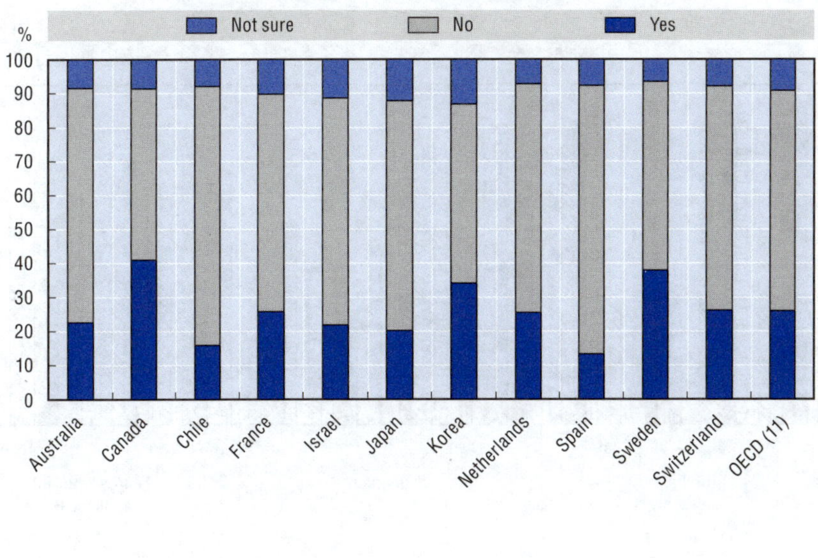

a detached house, income level, the environmental concern index and membership in an environmental organisation increases this probability of considering energy costs.

Energy efficiency investments

Respondents were requested to indicate whether they had invested in a number of energy-saving types of equipment (see Q77 in Annex A).[22] However, as with energy-saving behaviour, it is important to note that for many investments the motivations are likely to be a mix of private and public motives.

It is commonly found that owner-occupiers are more likely to invest in energy conservation equipment than tenants since they are able to recover the full benefits of the investment. This is likely to be particularly true of long-lived investments. However, differences in the reported behaviour are not large across the owner/not owner categories, except in the case of thermal insulation (see Figure 3.14). In fact, the largest relative difference is for "heat thermostats", where owner-occupiers are more likely to have installed such devices. However, this may not be purely a reflection of incentives. For instance, tenants may not be allowed to make such investments.

It is revealing that those who do not own their residence seem (marginally) more likely to have installed energy-efficient light bulbs. Since the "life" of bulbs is less than other types of capital equipment, and they can be easily transferred, split incentives are likely to be less important in this case. Other

Figure 3.14. **Reported energy-saving investment by home ownership status**

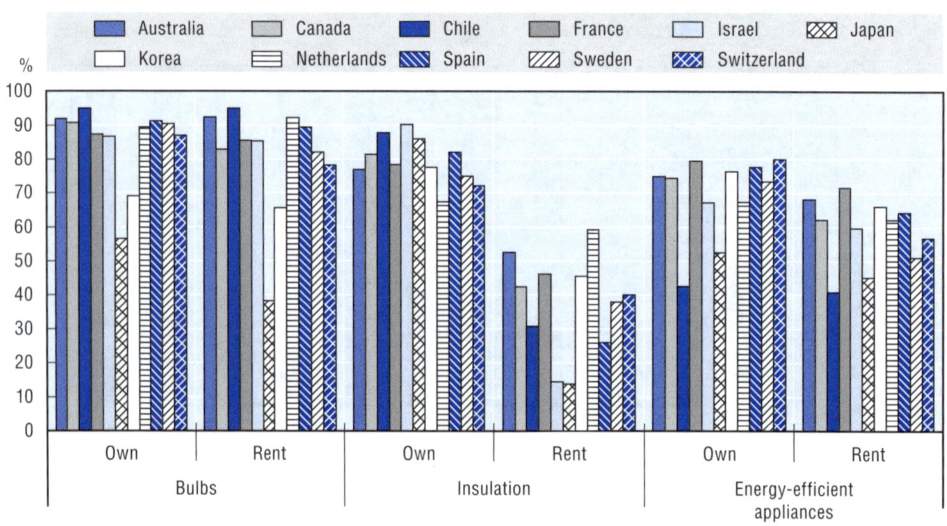

factors may however play a role. It turns out that those who have bought low-energy lamps tend to score higher on the energy behavioural index, have a slightly higher median income, are more likely to be members of an environmental organisation, and score higher on the environmental concern.

In the introduction to this chapter, some questions were raised about who invests in energy efficiency and how effective certain subsidies programmes are. It is quite possible that subsidies for investment in energy conservation equipment will actually lead to an increase in energy use. However, the main purpose here is to describe the usage of such programmes, and not to assess whether or not they have led to an overall decrease. Respondents were requested to indicate for a number of investments whether the household had benefited from government (or utility company) financial support (such as grants or preferential loans) (see Q78 in Annex A). In total, on 4 026 occasions, households have benefited from financial support out of a total of 25 737 installations. So, roughly speaking, the average household has installed two of the energy-saving items listed in the questionnaire. Assuming that each support measure is unique, it can be said that every sixth installation receives support.

More environmentally sensitive respondents are more likely to have benefited from support. This raises questions about the efficiency of the programmes. Equity concerns are also often raised when assessing the allocation of grants. The evidence indicates a slight progressivity in the allocation of grants, which is more pronounced in the case of insulation and much less so in the case of double- or triple-glazed windows (Figure 3.15).

Figure 3.15. **Reported receipt of energy efficiency grants, by income category**

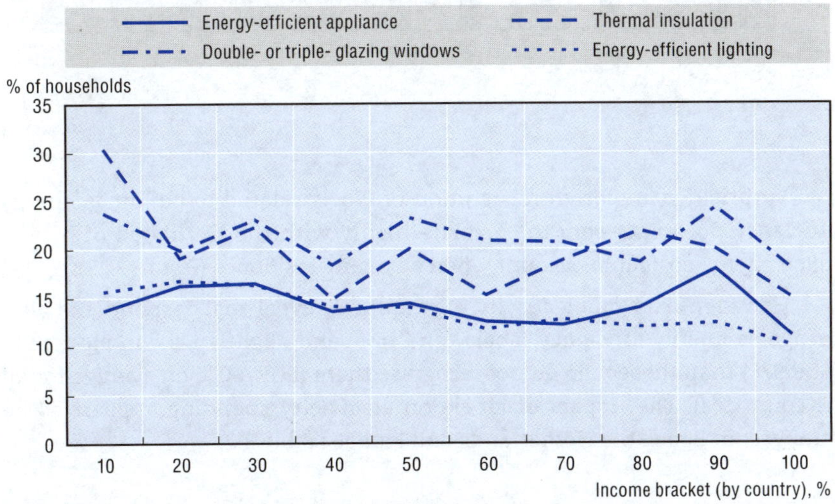

Note that brackets, representing deciles, differ by country. "Don't know" and "Prefer not to answer" represent approximately 15% of the total sample, and are excluded here.

The role of energy efficiency labels

Responses indicate a high level of recognition of labels for energy-efficient appliances in all countries (Figure 3.16). As in the 2008 survey, the level of recognition is highest in Australia (96%). Chile, France, Korea, the Netherlands and Switzerland are above the average (70%) while Spain and Sweden rank the lowest. By contrast, respondents are much less familiar with energy efficiency labels for buildings as Figure 3.16 shows for the seven countries which have introduced such labels.[23] One can note the almost reverse situation in Australia, which has the lowest level of recognition, followed by Sweden and the Netherlands. The energy efficiency label for buildings appears to be the best known in France and Korea.

Figure 3.16. **Recognition of energy efficiency label: appliances and buildings**

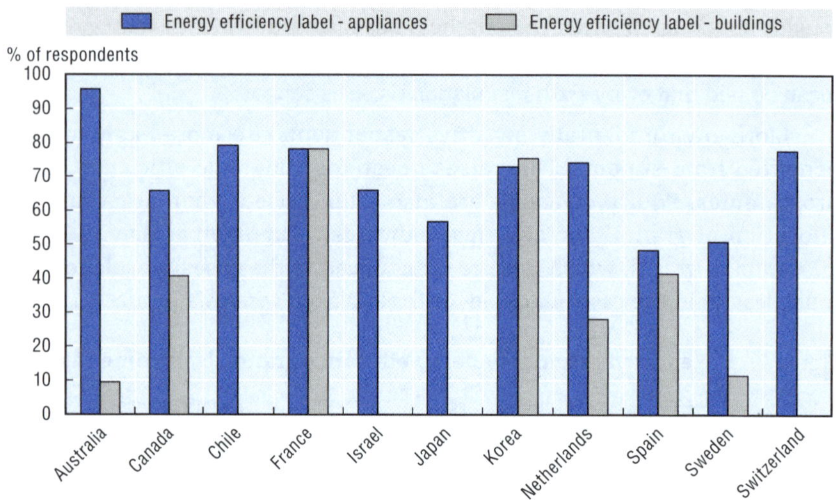

Surprisingly, reported recognition of labels related to buildings energy efficiency does not seem to correlate highly with the likelihood of having taken energy costs into account when changing residence (Figure 3.17).

However, data on electricity expenditure reveal that respondents who recognise energy efficiency labels tend to spend slightly less on electricity (EUR 976) than those who do not recognise them (EUR 1 036 on average for all 11 countries). The impact of labels on electricity spending appears to be strongest in Israel, the Netherlands and France (Table 3.2).

Figure 3.17. **Respondents reporting taking into account energy costs when changing home**

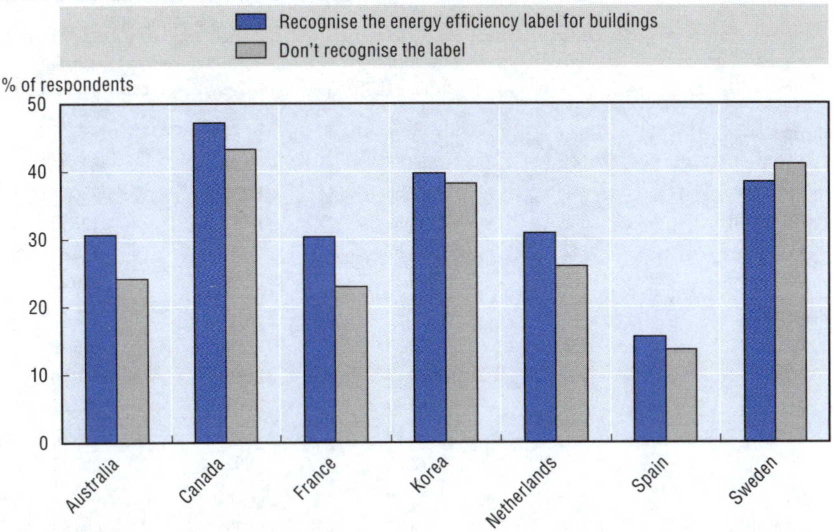

Motivations to reduce energy use

Respondents were asked to indicate the level of importance of a range of factors in their motivation to reduce their energy consumption. The results are summarised in Figure 3.18 which reveals remarkably consistent numbers across countries in several dimensions.

First, the price of energy-efficient equipment appears to be a relatively important factor; indeed, it is ranked highest in almost all countries, as in the 2008 survey. Conversely, respondents consistently rank energy price the lowest in their assessment of factors that would reduce their energy consumption. There may, of course, be some strategic bias in this apparent contradiction. However, it may also reflect the relatively short-term view they adopt when responding to the question, as well as the available evidence indicating a relatively low short-run price elasticity. Only marginal adjustments may occur in the first instance, when households are locked in their current stock of appliances, and have deep-seated behavioural patterns. This is also in line with the significant bulk of evidence on the high rate of return households seem to require before they make investments to save energy.

There are significant country variations in the replies. Spain, Chile, Israel, Korea and Spain are the countries where the provision of more practical information on how to reduce energy consumption at home and the easier identification of energy efficiency appliances are ranked the highest. Being more informed on the environmental impacts of energy consumption does

Table 3.2. **Reported annual expenditures on electricity, by recognition of energy efficiency labels**

	Don't recognise label		Recognise label	
	N	Electricity expenditure (EUR)	N	Electricity expenditures (EUR)
House size				
< 50 m	141	663.65	389	643.11
50-100 m	809	847.85	1 894	796.08
101-200 m	568	1 245.99	1 752	1 127.72
> 200 m	142	1 674.87	477	1 410.57
Don't know	75	969.36	261	960.20
Household size				
1	288	720.89	638	656.29
2	576	1 020.13	1 569	910.96
3	365	1 043.20	1 003	993.37
4	314	1 140.15	954	1 097.73
5	192	1 373.71	609	1 257.11
Income				
1st quartile	479	779.32	1 000	684.26
2nd quartile	416	1 028.98	1 089	921.46
3rd quartile	344	1 251.90	1 100	1 076.00
4th quartile	311	1 263.21	1 141	1 190.99
Missing	185	934.38	443	963.70
Country				
Australia	21	1 009.63	631	1 110.38
Canada	165	1 115.03	429	1 068.08
Chile	137	534.48	561	526.27
France	157	916.90	644	843.15
Israel	213	1 280.16	533	1 155.55
Japan	237	1 094.75	335	1 064.26
Korea	13	396.06	36	452.74
Netherlands	102	1 240.75	380	1 157.98
Spain	310	829.40	367	831.17
Sweden	256	1 424.22	306	1 485.34
Switzerland	124	725.05	551	856.20
Overall	1 735	1 036.16	4 773	975.73

not appear as an important factor motivating households to reduce energy consumption. This is particularly true of the Dutch and the Swedish respondents. A closer look at the results suggests that women tend to give higher scores for the different motivating factors, as well as members of an environmental organisation and older respondents. There are also variations across countries to be noted with Chilean respondents typically reporting a higher score than the average, and Dutch respondents a lower score.

Figure 3.18. **Motivations to reduce energy consumption at home, by country**

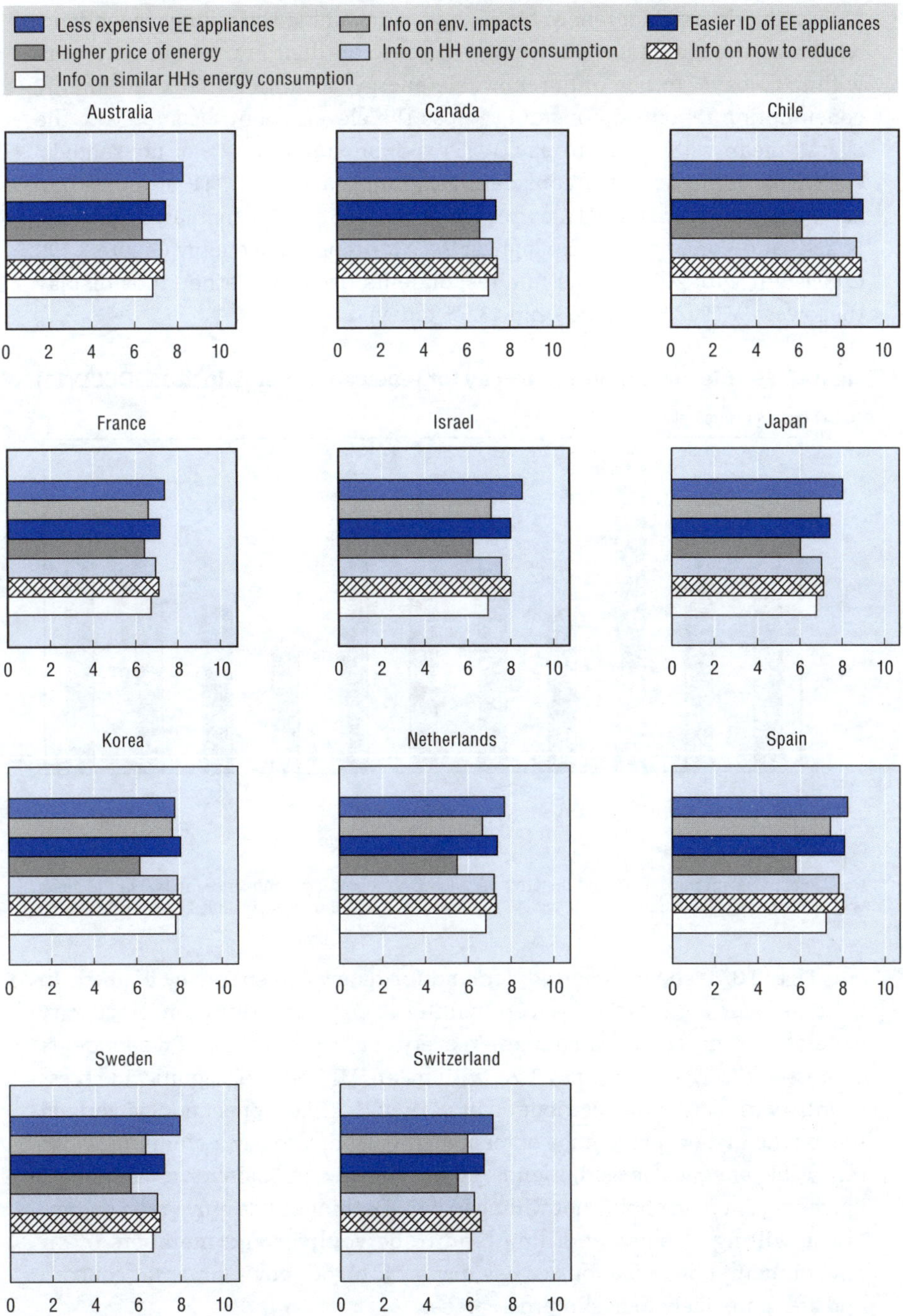

Note: Reported level of importance (0 = not important, 10 = very important).

5. Willingness-to-pay to use renewable energy

Latent demand for renewable energy was examined by asking respondents what was the maximum percentage increase on their annual bill they were willing to pay[24] to use only renewable energy, assuming that their energy consumption remained constant. Across the eleven countries surveyed, the overall mean willingness-to-pay (WTP) is approximately 12% of households' electricity bill. When only considering the replies with a WTP of more than zero, the overall mean WTP corresponds to about a 19% increase. Chileans, Israelis and Koreans state the highest WTP for renewable energy (Figure 3.19). Consistent with 2008 results, the respondents from the Netherlands display the lowest WTP (7.5% in 2008 and 13.5% today).

Figure 3.19. **Mean willingness-to-pay for renewable energy in 2011, OECD(11)**

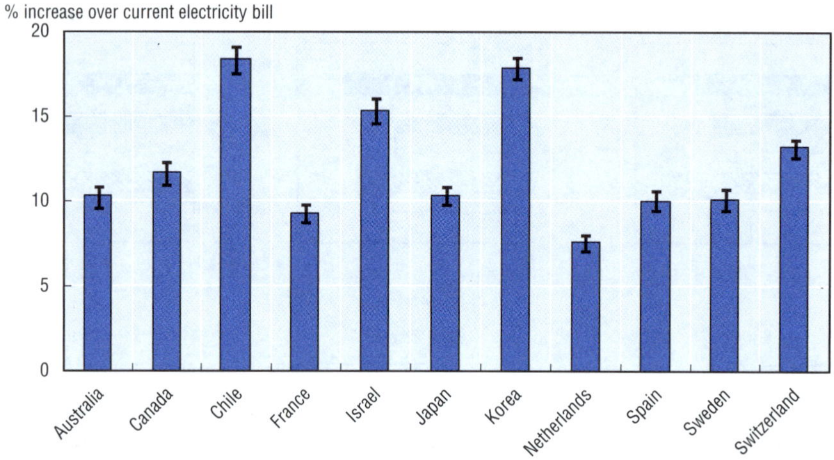

Note: Bars represent mean WTP and the error bars above represent 95% confidence interval of the mean WTP. Note that all respondents – those with WTP = 0 and with WTP > 0 – were included in this calculation.

Figure 3.20 shows reported demand for renewable energy by country. In general, over 60% said they were willing to pay something for electricity generated from renewable sources. In every country, over half of respondents expressed a WTP greater than 0. As with mean WTP, there is significant cross-country variation in the distribution of WTP, with Chile, Korea and Switzerland having the highest percentage of respondents willing to pay a premium to use renewable energy. The Netherlands, Japan, France and Spain have the highest percentage of households not willing to pay anything to use only green energy. Those willing to pay something tend to be younger and members of an environmental organisation.[25] They display a higher environmental concern and are more likely to live in urban areas.

Figure 3.20. **Willingness-to-pay for renewable energy, by country**

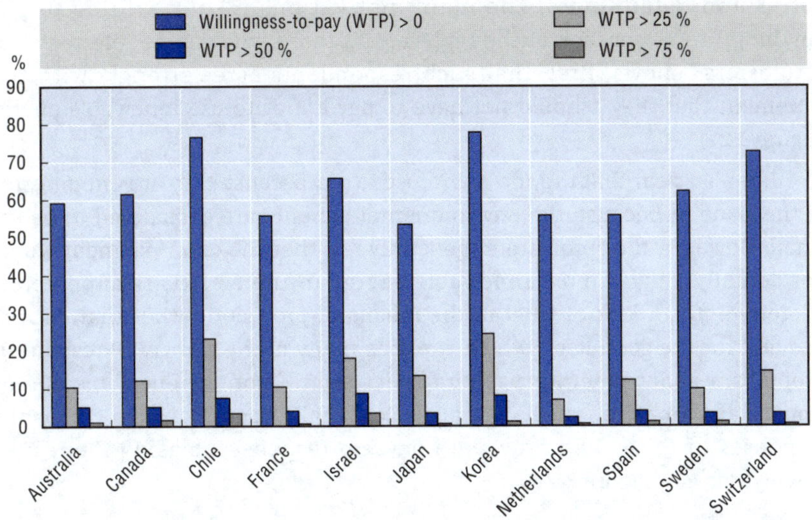

While one could expect to find that WTP is comparatively higher in countries that have invested relatively less in renewable energy,[26] the results indicate no obvious relationship at this stage between WTP and the share of renewables in a country energy mix. One can just note that Korea has a low share of renewables and a relatively high WTP*[27] (18%), a finding in line with Kwak and Yoo (2009).

The mean values are considerably higher than those found – for a different sample, of course – in the first round of the survey. However, this may be partly attributable to the more fine-grained nature of the data collected in the second round. More specifically, in the former case respondents had to select among five brackets, while in the latter, a slider was used with potential values for any integer.[28]

Moreover, the median from the 2011 sample is close to the value observed in the 2008 survey. When restricting the comparison to the countries in both surveys[29], the very same picture emerges; the means appear to be higher in the 2011 survey, but the medians are quite similar. Furthermore, in both surveys, about 6 in 10 respondents are willing to pay something to use renewable energy. Note that the median is lower than the mean in both cases.

A follow-up question was added to the 2011 questionnaire to have a better understanding of the main motivations for not willing to pay anything.[30] A closer look at the replies from those who can be referred to as "protesters" shows, first, that such respondents agree strongly with the statement that they "should not have to pay extra" to use renewable energy (Figure 3.21).

It is also clear that they do not refuse to pay because they are uninterested in the issue or because the environmental benefits are considered to be too small. However, the "protesters" generally felt that the extra payment would not actually deliver more renewable energy in the electricity supply mix. Therefore, it is a lack of trust in the means of provision rather than lack of demand for the good itself which explains many of the zero-WTP responses. Scepticism about whether paying for extra provision of renewable energy would have an effect on the supply mix actually used may also colour the responses of those who provided non-zero responses, biasing the results downwards (see Figure 3.22).

Figure 3.21. **Reasons for not wanting to pay any extra for renewable energy**

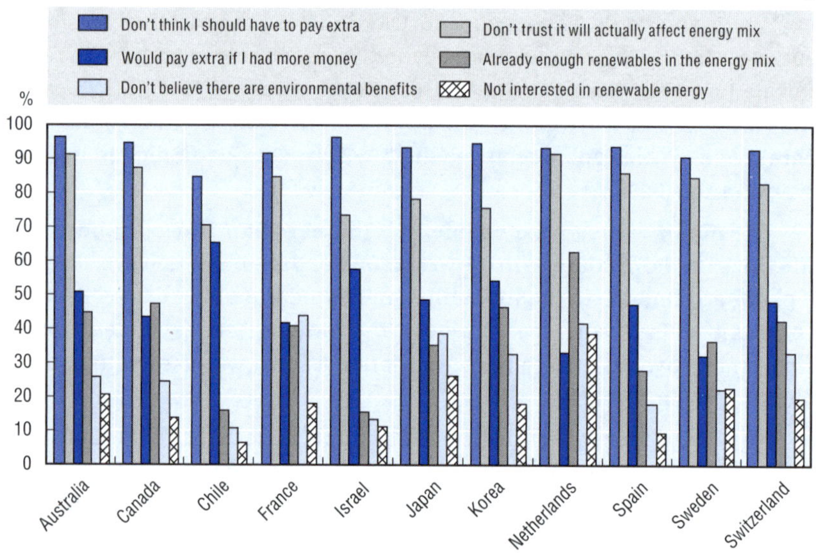

The data confirm the strong influence of households' attitudes towards the environment on the willingness-to-pay to use renewable energy. This result is consistent with some previous literature.[31] In particular, there is evidence in all countries that membership in environmental organisations correlates positively with WTP for renewable energy. The overall WTP* (including WTP = 0) is a little more than 11% and reaches almost 19% when the

Figure 3.22. **Willingness-to-pay for renewable energy by level of trust in environmental information from government**

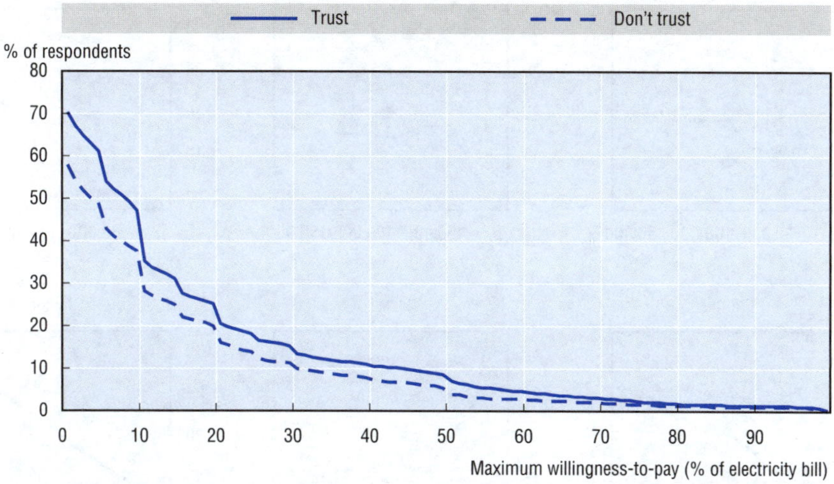

Note: This figure presents a survival function showing the proportion of respondents with a WTP for renewable energy greater than a given percentage. As the percentage increases, the proportion of respondents with WTP exceeding this percentage decreases. Demand for renewable energy is presented for full sample here while Figure 3.20 shows results by country.

person is a member of an environmental organisation. For some countries (e.g. Sweden and Australia) there are particularly significant differences between WTP values for those who are members of environmental organisations and those who are not.[32] These disparities are much smaller if the focus is on those who are willing to pay more than zero – i.e. not being a member correlates highly with zero-WTP responses.

Respondents were also asked if they were "willing to make compromises in [their] current lifestyle for the benefit of the environment" (see Q26 in Annex A). Comparing these responses with WTP values reveals internal consistency in the data. Those who "strongly agreed" with this statement report an average WTP which is more than twice that of those who strongly disagreed with this statement (16.5% vs. 7.5%).

Casual inspection of the descriptive data does not reveal a significant relationship between income and WTP. However, there is some weak evidence of a non-linear relationship. In general, the value of WTP tends to be higher at the lower-income levels, then decreases to a certain level of income, after which WTP increases with income. However, this relationship differs by country (see Figure 3.23).

Figure 3.23. **Relationship between willingness-to-pay and income level**

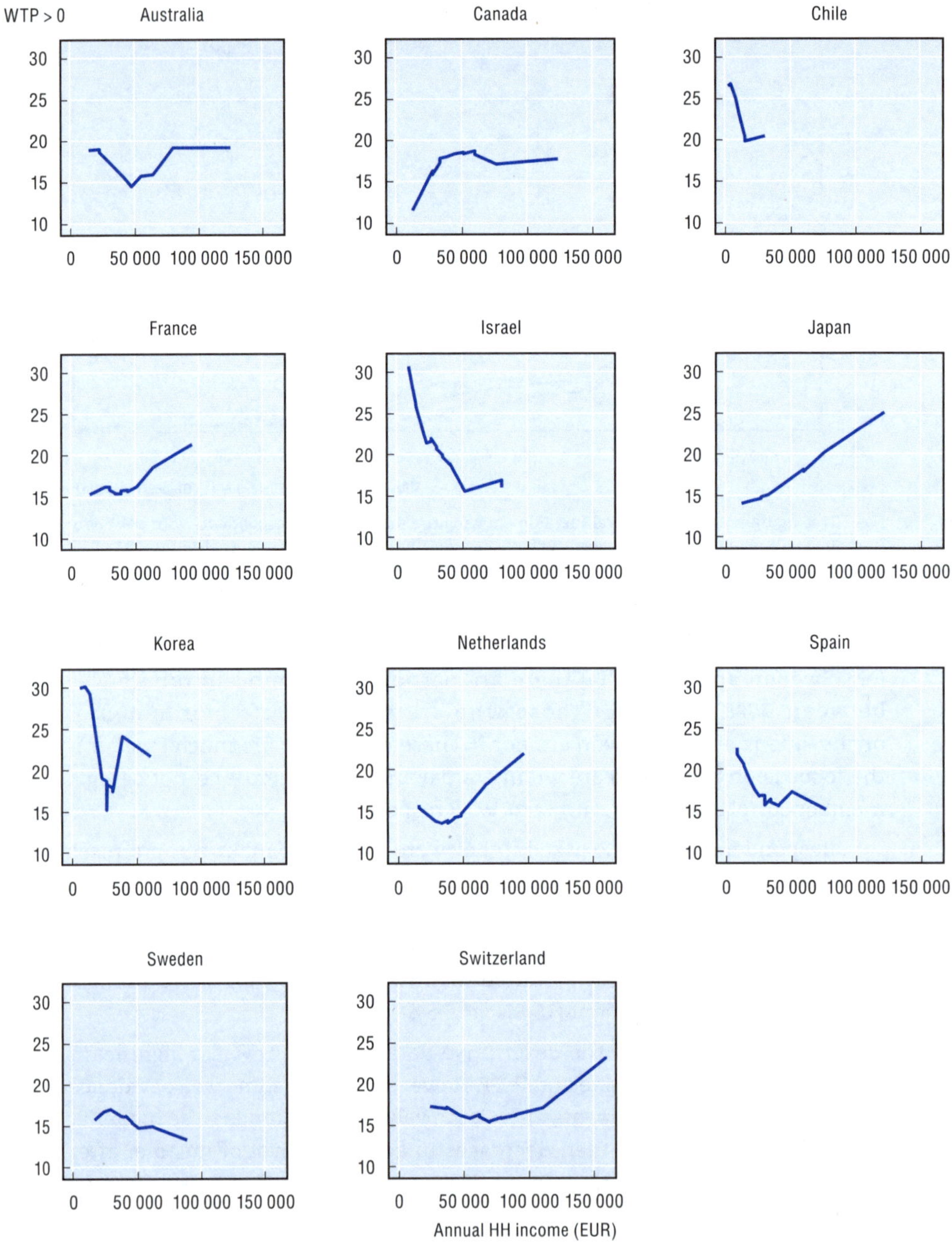

Note: The data are displayed as the predicted values from Lowess-smoothed estimation.

6. Conclusions

The main findings arising from the descriptive analysis of survey responses relate to energy-saving behaviour and the demand for renewable energy can be summarised as follows. First of all, a comparison with the previous EPIC Survey implemented in 2008 shows that some of the results are, indeed, remarkably consistent. In particular, households' attitudes towards the environment appear to be a driver in various decisions involving renewable energy and energy-saving behaviour (see OECD, 2011).

Electricity spending patterns: Respondents report an average yearly spending on electricity of about EUR 960, the average budget share being about 3% at a daily consumption of 18 kWh (range 9 to 34 kWh). These numbers are consistent with similar surveys.

Electricity consumption: The price elasticity of demand for electricity is less than one and in line with previous analyses. More than 90% of the respondents report residential electricity metering. This figure is almost exactly the same as in the previous survey. The data support an income elasticity of demand for electricity less than one in each country. This adds to the bulk of evidence suggesting that higher energy prices have regressive effects.

Demand for renewable energy: The overall mean WTP corresponds to about a 12 % increase of the electricity bill. In both the 2008 and 2011 survey, about 6 in 10 respondents are willing to pay something to use only renewable energy. Consistent with the first survey, respondents from the Netherlands display the lowest WTP. There is a significant difference between the mean and the median in the new survey. The income elasticity of demand for renewable energy is found to be less than one. The findings confirm the role of attitudes towards the environment on the WTP for renewable energy.

Energy-saving behaviour: A majority of the respondents, in each country, claim to "always turn off the light when leaving a room" and are engaged in other forms of energy-saving behaviour. The fact that roughly 40% of the respondents "occasionally" or "never" completely turn off their stand-by appliances suggests that there are important potentials for energy conservation in most countries. In terms of energy-saving behaviour, Canada, Japan, the Netherlands, Switzerland and Sweden score lower than the average while Spain ranks at the top. Women and home-owners are more likely to undertake energy-saving activities. Results also suggest that income has a negative impact on energy-saving behaviour, and confirm the positive effect of concerns for the environment.

Investments and the use of government support schemes: The price of energy-efficient equipment is the key factor in all countries; it is ranked highest in almost all countries (among seven factors thought to encourage households to reduce their energy consumption). Conversely, respondents typically rank increased energy prices lowest in their assessment of factors that would encourage them to reduce their energy consumption. This is certainly largely a consequence of a strategic bias in responses. Households commonly benefit from government support for investments in renewable energy and energy efficiency. Approximately 4 000 of 25 000 installation of measures have received support. Households do not commonly invest in renewable energy technologies like wind and solar. A closer look at the data shows that Israeli households, predictably, are very far ahead regarding the instalment of solar panels.

The role of labelling: The analysis suggests that energy efficiency labelling may have a small effect on electricity spending. Recognising the relevant labels correlates with a lower level of consumption. However, there is little apparent relationship between recognition of labels for buildings' energy efficiency and the likelihood of taking energy costs into account when moving residence.

Overall, in line with the first survey (OECD, 2011), these findings clearly confirm the significant role of attitudes and norms in shaping the energy-related behaviour underlined in recent literature, implying that policy measures based on the provision of information to consumers and on education can usefully complement incentive-based measures.

This first overview of the survey data offers food for thought on energy use. The descriptive analysis will be refined in a second publication where formal statistical tests will be carried out to test some hypotheses and draw policy implications.

Notes

1. While the literature on residential energy demand focuses mostly on economic variables, the role of attitudes, norms and other "non-economic" variables has received increased attention (see Martinsson et al., 2011; Dowlatabadi and Wilson, 2007).
2. In 2008. See Table 2.14 in IEA/OECD (2010).
3. For a recent survey of the links between electricity consumption and economic growth, see Ilhan and Ozturk (2010).
4. See Table 3.7 in IEA/OECD (2010).
5. See EnergyAustralia's residential customer price list. *http://bit.ly/rkx7JE*.
6. See Sonderegger (1977) for an early paper on this line.
7. Haneda et al. (1996) discuss how modes of space heating in Japan and Scandinavia are due both to climatic differences, but also to social norms.

8. As per the Jevons paradox, *technological progress* that increases the *efficiency* with which a *resource* is used tends to increase (rather than decrease) the rate of *consumption* of that resource. It is also referred to as the "rebound effect" for improved energy efficiency.
9. Dubin and McFadden (1984) make these points. Theirs is a key paper that sets the stage for a large number of econometric analyses of residential energy demand.
10. See Cayla et al. (2011) for a recent assessment.
11. The percentage was almost exactly the same in the 2008 OECD survey (90.4%) for a different set of countries.
12. See for example *http://bit.ly/cXawF7*.
13. See Table 3.7 in IEA/OECD (2010).
14. See for example EnergyAustralia's new regulated retail tariffs and charges for residential customers for 2011, *http://bit.ly/rkx7JE*.
15. Respondents who indicated they "don't know" are excluded from the sample. In the Swedish case, a bill was passed in 2012 (after the survey was implemented) enforcing hourly metering at no extra cost for those consumers subscribing to hourly-based electricity supply contracts.
16. However, it should be noted that virtually all technologies for generating electricity may cause some environmental problems.
17. The possible answers being "probably true", "definitely true", "probably not true", "definitely not true" and "don't know".
18. For the record, the average consumption is about 700 kWh lower per year in that group. Thus, there is at least a weak indication that smart meters have an effect on consumption.
19. Question Q7a in ECME (2011, p. 203).
20. For example, the European Commission writes in its Regulation (EC) No 1275/2008 of 17 December 2008 (*http://bit.ly/qZ7NkQ*): "It has been stated in the preparatory study that stand-by functionalities and off-mode losses occur for the majority of electrical and electronic household and office equipment products sold in the Community, while the annual electricity consumption related to stand-by functionalities and off-mode losses in the Community has been estimated to be 47 TWh in 2005, corresponding to 19 million tonnes (Mt) of CO_2 emissions. Without taking specific measures, consumption is predicted to increase to 49 TWh in 2020. It has been concluded that the electricity consumption of stand-by functionalities and off-mode losses can be significantly reduced". The regulation requires that "Power consumption of equipment in any off-mode condition shall not exceed one watt" with certain amendments.
21. These two countries can be considered as "colder" than the others, which can also be an explanatory factor.
22. They could indicate yes, no, already equipped, not possible.
23. An energy efficiency label for buildings was shown to respondents in Australia, Canada, France, Korea, the Netherlands, Spain and Sweden.
24. The willingness-to-pay (WTP) is the maximum amount a person would be willing to pay, sacrifice or exchange in order to receive a *good* or to avoid *something undesired*, such as pollution.

25. See Parker et al. (2003).
26. Assuming that the marginal utility of environmental improvements is increasing at a falling rate.
27. WTP* designates WTP that includes the zero values (WTP=0). In other cases, the zero values are removed from the sample.
28. Moreover, the structure of the question differs somewhat between the two rounds.
29. The six countries that took part in the two rounds of the survey are: Australia, Canada, France, Korea, the Netherlands and Sweden.
30. Respondents who displayed a WTP=0 were asked "would you not be willing to pay more to use only renewable energy?" and had to indicate how much they agreed with six different statements (Q.72).
31. See for instance Gamble et al. (2008) who found that WTP for electricity among Swedish households increased with attitude towards the environment.
32. Membership varies in the countries from 3% in Japan to 20% in Switzerland, the average being about 12%.

References

Allysa, F. et al. (2007), *The twin pillars of sustainable energy: Synergies between energy efficiency and renewable energy technology and policy*, Technical Report E074, American Council for an energy efficient economy.

Abrahamse, W. and L. Steg (2009), "How do socio-demographic and psychological factors relate to households' direct and indirect energy use and savings?", *Journal of Environmental Psychology*,Vol. 30, pp. 711-720.

Abrahamse, W. et al. (2005), "A review of intervention studies aimed at household energy conservation", *Journal of Environmental Psychology*,Vol. 2, No. 3, pp. 273-291.

Barr, S., A. Gilg and N. Ford (2005), "The household energy gap: examining the divide between habitual and purchase-related conservation behaviors", *Energy Policy*, No. 33, pp. 1425-1444.

Baumol, W. and E. Wolff (1981), "Subsidies to new energy sources: Do they add to energy stocks?", *Journal of Political Economy*,Vol. 89, pp. 891-913.

Brandon, C. and A. Lewis (1999), "Reducing household energy consumption: a qualitative and quantitative field study", *Journal of environmental Psychology*,Vol. 19, pp. 93-112.

Carlsson-Kanyama, A. and A. Lindén (2007), "Energy efficiency in residences – challenges for women and men in the North", *Energy Policy*, Vol. 35, pp. 2163-2172.

Cayla, J., N. Maizi and C. Marchand (2011), "The role of income in energy consumption behaviour: Evidence from French households data", *Energy Policy*, Vol. 39, pp. 7874-7883.

Dillman, D., E. Rosa and J. Dillman (1983), "Lifestyle and home energy conservation in the United States: the poor accept lifestyle cutbacks while the wealthy invest in conservation", *Journal of Economic Psychology*,Vol. 3, pp. 299-315.

Dowlatabadi, H. and C. Wilson (2007), "Models of decision making and residential energy use", *Annual Review of Environment and Resources*,Vol. 32, pp. 169-203.

Dubin, J. and D. McFadden (1984), "An econometric analysis of residential appliance holdings and consumption", *Econometrica*,Vol. 52, pp. 345-362.

ECME (2011), *The functioning of the retail electricity markets for consumers in the European Union*, Technical Report EAHC/FWC/2009 86 01, Directorate-General for Health and Consumers, European Commission.

Firestone, R. et al. (2011), *The time for (behavior) change is now: Applying social marketing principles to residential energy efficiency programs*, Technical report, Navigant Consulting.

Gamble, A. et al. (2008), "Psychological determinants of attitude towards and willingness to pay for green electricity". *Energy Policy*,Vol. 36, pp. 768-774.

Gustavsson, L., G. Nair and K. Mahapatra (2010), "Factors influencing energy efficiency investments in existing Swedish residential buildings", *Energy Policy*,Vol. 38, pp. 2956-2963.

Hilhite, H. et al. (1996), "A cross-cultural analysis of household energy use behaviour in Japan and Norway", *Energy Policy*,Vol. 24, No. 9, pp. 795-803.

Hedberg, P. and S. Holmberg (2005), *Saving energy*, SOM-Institute.

IEA/OECD (2010). *Electricity Information 2010*, International Energy Agency, 9 rue de la Fédération, 75739 Paris Cedex 15, France.

Ozturk, I. (2010), "A literature survey on energy-growth nexus", *Energy Policy*, Vol. 38, No. 1, pp. 340-349.

Kriström, B. and P. Riera (1996), "Is the income elasticity of environmental improvements less than one?", *Environmental and Resource Economics*,Vol. 7, pp. 45-55.

Kwak, S.-Y. and S.-H. Yoo (2009), "Willingness to pay for green electricity in Korea: A contingent valuation study", *Energy Policy*,Vol. 37, pp. 5408-5416.

Marechal, K. (2010), "Not irrational but habitual: The importance of behavioural lock-in in energy consumption", *Ecological Economics*,Vol. 69, No. 5, pp. 1104-1114.

Martinsson, J., L. Lundquist and A. Sundström (2011), "Energy saving in Swedish households. The (relative) importance of environmental attitudes", *Energy Policy*,Vol. 39, pp. 5182-5191.

Menges, R., C. Schroeder and S. Traub (2005), "Altruism, warm glow and the willingness-to-donate for green electricity: An artefactual field experiment", *Environmental and Resource Economics*,Vol. 31, pp. 43-458.

Mosak, J. (1938), "Interrelations of production, price, and derived demand", *Journal of Political Economy*,Vol. 46, No. 6, pp. 761-787.

OECD (2008). "Residential Energy Demand, by Bengt Kriström", *Household Behaviour and Environmental Policy: Reviewing the Evidence*, OECD Publishing.

OECD (2011), Greening Household Behaviour: The Role of Public Policy, OECD Publishing. doi: *10.1787/9789264096875-en*.

Parker, P., I. Rowlands and D. Scott (2003), "Consumers and green electricity: profiling potential purchasers", *Business Strategy and the Environment*, Vol. 12, pp. 36-48.

REN21 (2011), *Renewables 2011 global status report*, Technical report, Renewable Energy Policy Network for the 1st Century (REN21) Secretariat. Available from *www.ren21.net/gsr*.

Sav, G. (1983), "On subsidies for energy-saving durables", *The American Economist*, Vol. 30, No. 1, pp. 56-59.

Sonderegger, R. (1977), "Movers and stayers: the resident's contribution to variation across houses in energy consumption for space heating", *Energy and Buildings*,Vol. 1, pp. 13-324.

Stead, D. (2009), "Energy efficiency: Temporal and geographical trends in Europe", *Energy & Environment*, Vol. 20, pp. 345-365.

Stern, P. (1992), "What psychology knows about energy conservation", *American Psychologist*,Vol. 47, No. 10, pp. 1224-1232.

Greening Household Behaviour
Overview from the 2011 Survey

Chapter 4

Household behaviour and transport choices

by
Claude Weis *and* Kay W. Axhausen*

This chapter presents an overview of the data on the determinants of households' personal transport choices. It examines the effects of various types of public policies influencing transport demand, such as financial incentives to buy "cleaner" vehicles or car labelling. It also looks at differences in behaviour across households and the effect of norms and attitudes.

* Institute for Transport Planning and Systems, Swiss Federal Institute of Technology, Zurich, Switzerland.

The statistical data for Israel are supplied by and under the responsibility of the relevant Israeli authorities. The use of such data by the OECD is without prejudice to the status of the Golan Heights, East Jerusalem and Israeli settlements in the West Bank under the terms of international law.

1. Introduction

Societies bear substantial environmental costs from individuals' local and regional transportation choices. At the local level, traffic congestion in urban areas decreases air quality and hinders economic productivity and well-being. Globally, the transport sector as a whole currently accounts for around 30% of CO_2 emissions from fossil fuel combustion (OECD, 2010). Within Europe, traffic-related air pollution has been estimated to be responsible for approximately 3% of total yearly mortality (Chanel et al., 2000).

The social costs of transport choices are rarely borne directly by the individuals making these choices, and it is this fact that has led economists and policy makers to propose policy instruments which in some way pass these costs (or the benefits of eliminating them) on to travellers. Such incentives – including, for example, congestion charging schemes (Jonas, 2009; Bocarejo and Prud'homme, 2005), low-emission zones (Beevers and Carslaw, 2002), vehicle scrapping subsidies and rebates (Mian and Sufi, 2010), improved parking systems (Axhausen et al., 2011), municipal bike-share programmes (Dill et al., 2010), and the European Union's recently imposed carbon tax on airlines (Meltzer, 2012) – have rightly been the focus of most economic research on transport policy.

Yet, in addition to economic incentives, individuals' attitudes towards the environment and public policy can affect vehicle purchase decisions, propensity to use public transportation, and support for government policies addressing environmental impacts of transport systems. This chapter examines the role these attitudes play in shaping transportation choices among households sampled in the 2011 Survey on Environmental Policy and Individual Behaviour Change (EPIC). Key questions addressed here are:

- How is the decision to purchase a car – and what type to purchase – linked to *a)* environmental considerations and *b)* the presence of different public transportation options?
- How do attitudes towards the environment correlate with the use of private cars, public transportation and other modes of transport?
- What characteristics of public transportation systems are most important to households in considering their use?
- Which types of households support government policies to limit greenhouse gas emissions (GHG) from motor vehicles?

- Do car labelling schemes affect households' purchasing decisions?
- How much are households willing to pay (WTP) for an electric car, and what is the perceived value of these cars to different socio-economic groups?

To address these questions, a summary of relevant data and previous research on these topics is first provided, before turning to a presentation of the transportation-related data from the survey. Data on car ownership, car use, choice of travel mode for frequent trips are presented and also on respondents' support for different types of government policies aimed to reduce the environmental impacts of transport systems. To conclude, preliminary policy lessons from the analysis are highlighted which point out where further research in this area would be most valuable. A summary of key findings on transport choices is provided in Box 4.1.

Box 4.1. **Personal transport choices: key findings**

Findings from descriptive analysis suggest that:

- Across countries, there is broad household support for additional government investment in public transportation infrastructure. Households most frequently cited "improved public transport" as the factor which would lead them to use their cars less, and investment in public transport was the second most popular means of reducing vehicle CO_2 emissions (after subsidies for less polluting cars).
- Across countries, there is significant willingness to pay an additional price premium for the purchase of electric cars. Lack of infrastructure for such cars is cited by respondents as one of the main roadblocks to their uptake.
- Environmental attitudes play an important but subtle role in households' transportation choices. Households with a higher concern for environmental issues are more likely to use public transport and have a higher willingness to pay for electric cars.
- Car ownership is not strongly associated with environmental awareness and concerns. The relationship between environmental concerns and car use (given ownership) is much stronger, although it varies by country.

2. Overview

Transportation modes in the countries surveyed

Across the OECD, road travel by car or bus has consistently been both the dominant transportation mode and the largest transport-related contributor to CO_2 emissions (OECD, 2011b). Within the countries participating in the 2011 EPIC Survey, OECD statistics for 2008 indicate that, in general, the dominant mode of land-based motorised travel is by private cars, followed by rail, and lastly by bus (Figure 4.1). Exceptions to this trend can be found in Australia and Canada, where reported use of bus travel exceeds rail use and where private car use is highest.

This is to be expected, given the large geographic areas and relatively low population densities of these two countries. What is more striking is the reported intensity of private car use in France and Switzerland (given these countries' higher population densities) and, to some extent, the dominance of bus travel over rail in Spain (where rail systems are relatively well developed). On the basis of the survey responses, Japanese travellers are the most intensive users of their rail systems, while Spanish and Korean travellers are the most intensive users of buses and coaches. Comparable data for other modes of transport – such as walking and cycling – are difficult to find from secondary sources. Indeed, this is a key benefit of the 2011 EPIC Survey data described here.

Figure 4.1. **Reported use of land-based transportation by mode (2008)**

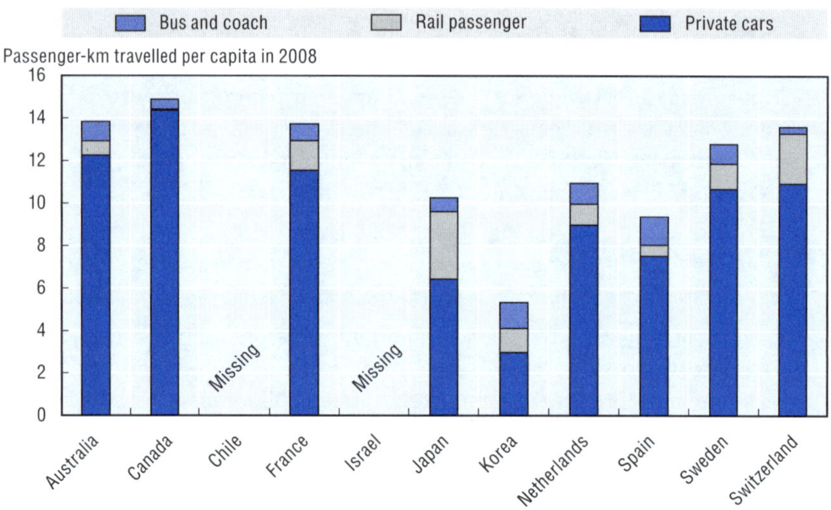

Previous research on environmental attitudes and transportation behaviour

(OECD, 2011a) provides a broad overview of the existing literature on the influence of certain personal characteristics on travel behaviour, particularly in terms of car ownership and use. Both variables are closely related, and appear to be influenced in similar ways by individual and household attributes. In the studies from the United States and Europe reviewed in (OECD, 2011a), both car ownership and use were found to be positively influenced by income and household size.

The effects of other personal characteristics, while also playing an important role in transport choices, can be more complex in nature and can vary from one region to another. For example, the influence of age on choice of transportation mode is in general non-linear. That is, private car use can be expected to increase with age early in life and then to decrease later in life (see Axhausen and Weis, 2009 for a model of the effects on general activity participation). Yet, many cultural factors can determine car and public transport use across different age groups. For example, elderly populations in some regions continue to drive into later ages more than in other regions. Such factors need to be carefully analysed in the context of each study before drawing conclusions about general trends.

With regard to policy factors influencing individual decisions, one would expect that enhanced accessibility to public transport leads households to use their cars less. This hypothesis has been confirmed in a few studies from Austria (Axhausen and Simma, 2004), the United Kingdom (Dargay and Hanly, 2004; Dargay and Giuliano, 2006) and the United States (Dargay and Giuliano, 2006). In the 2008 round of the EPIC Survey (OECD, 2011a), the same pattern in car ownership was found (higher accessibility to public transport was associated with lower car ownership). However, the opposite trend was found with car use (better access to public transport was associated with higher car use). The same survey data provided indications that awareness of environmental issues was associated with lower car use. In a recent study in the Swiss context (Axhausen et al., 2010), similar effects on mode choice were observed both with regard to the influence of socio-economic characteristics and particularly with the quality of the public transport supply. For reasons discussed in the opening chapters of this book, variables describing both attitudes and policy-related factors – and their interaction – are of particular interest.

3. Clustering households by their environmental concerns

There are many different ways of analysing how attitudinal data predict households' decisions. In the context of transportation, this chapter relies upon the use of cluster analysis for analysing the impact of a large number of attitudinal questions (Norusis, 2011). Such an analysis allocates respondents to a statistically determined number of clusters according to how they answered a set of attitudinal questions. Respondents giving similar answers across a range of questions are clustered together. This technique was applied in Chapter 2 as well. However, rather than focusing here on attitudes as reflected in the reported level of agreement with different statements, the focus is on responses to questions related to levels of environmental concern. More specifically, questions used for this procedure concerned the seriousness of six environmental issues. Recall that the response format for these questions was a 0- to 10-point scale, with 0 indicating "not at all serious" and 10 "extremely serious".

Figure 4.2 shows the mean of those scale variables for each of the three clusters created by this procedure. As can be seen, members of one cluster exhibit roughly the same average concern for all the indicator variables. In all clusters, the variation in valuation is greatest for the concern about biodiversity. These environmental concern clusters are used as summary variables in subsequent figures illustrating the association between environmental attitudes on travel behaviour. The cluster membership is used in subsequent sections of the chapter as a summary measure of environmental concern.

Throughout the discussion, it is important to recognise that environmental attitudes and concerns are correlated with other respondent characteristics, many of which – such as household location – are relevant for transport decisions. As shown in Figure 4.3, those in the "highly concerned" environmental cluster are more likely to live in urban areas. This is important because environmental awareness could be mistakenly concluded to directly affect transportation decisions, when in fact it is the household's location and the availability of transportation alternatives that is the influential factor in transportation decisions. Of course, there is the more fundamental question of how households' choice of location is linked to their environmental awareness: There is likely "residential self-selection", meaning that people with similar attitudes and cultures prefer to live near one another (Mokhtarian, 2008; Cao et al., 2009). While the data presented here are valuable for refining questions like this for further research, they cannot be used to answer this question directly. Indeed, the question of "residential self-selection" is one that is well studied in the academic community.

Figure 4.2. **Environmental concern clusters**

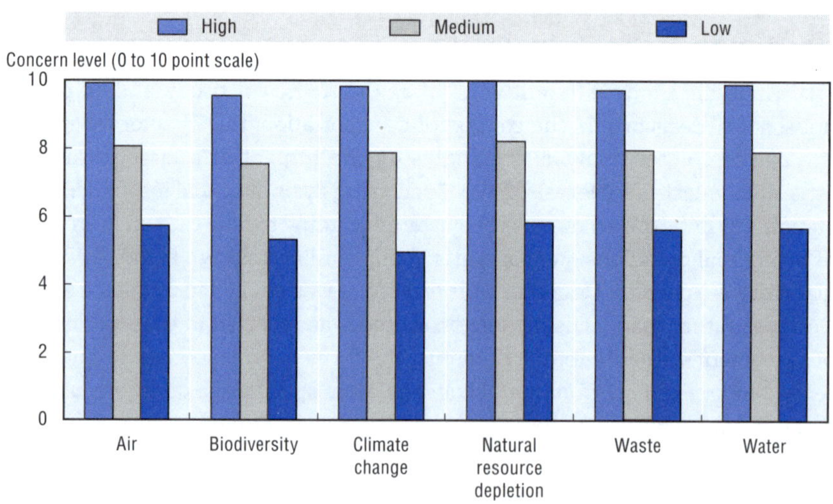

Figure 4.3. **Environmental concern and household location**

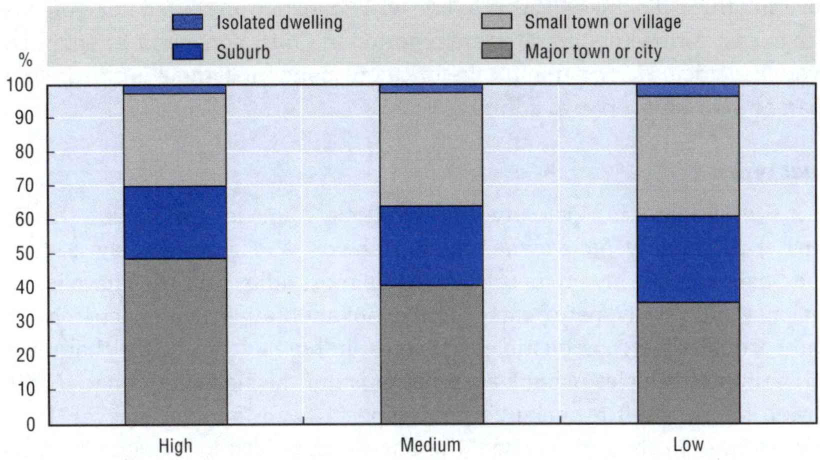

4. Car ownership

Respondents indicated car ownership rates which were slightly more than 2008 ownership rates reported by the World Bank (2008). Figure 4.4 shows the average values for the number of cars owned per household and per adult in the eleven participating countries, as well as in the whole sample. The 2008 OECD survey data (for those six countries that are common to both survey rounds) and World Bank data are included for comparison. Reported car ownership rates are among the highest in Australia and Canada, where travel

Figure 4.4. **Reported car ownership rates by country**

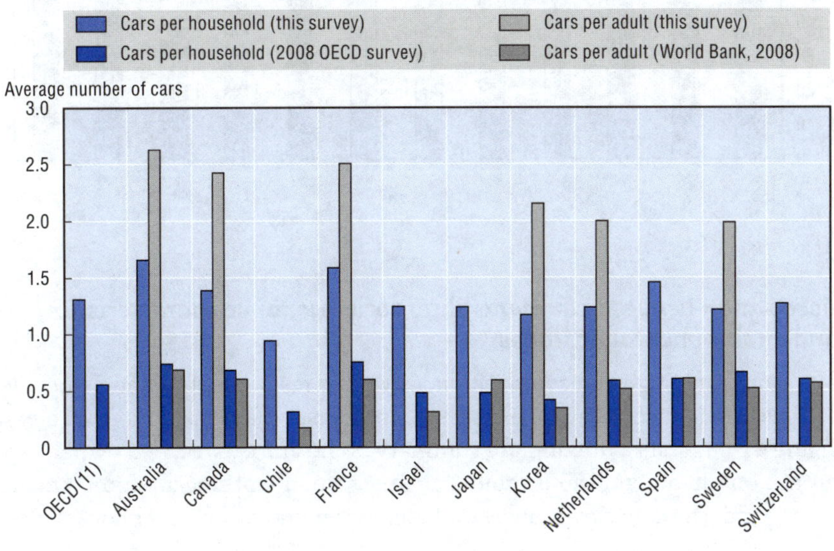

distances are traditionally long, and the car often the only viable option for covering those distances (this is in line with statistics presented in Figure 4.1). The lowest car ownership rates are reported in Chile, Korea and Sweden. The overall average car ownership rate for the respondent sample is a little more than one car for every two adults.

Fuel types

Vast differences exist between the eleven countries concerning the fuel type of the main vehicle owned by households. As Figure 4.5 shows, France and Spain rely heavily on diesel cars, while most other countries have a large majority of unleaded petrol cars. The share of alternative fuels (electric, hybrid and particularly biocombustibles) is largest in Korea. It is notable that stated ownership of vehicles using leaded petrol is non-negligible in Canada, Chile, Israel, Korea, the Netherlands, and especially in Japan. However, these statistics are likely to be revised: the phase-out of lead in gasoline has been under way for some time in these countries, so the statistics presented here on this aspect deserve scrutiny in future analysis.

Figure 4.5. **Reported distribution of fuel types among vehicle owners**

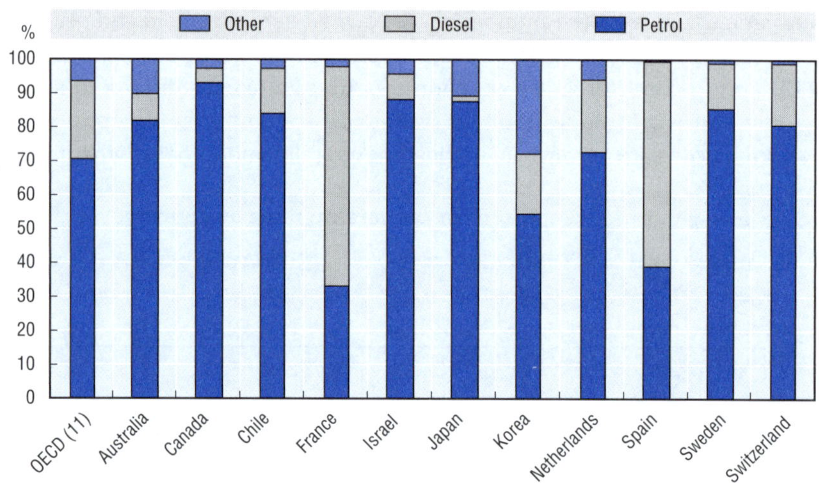

Association between car ownership, socio-economic characteristics, and environmental attitudes

Household size and composition, as well as respondents' gender, income and occupation, could all be expected to influence the decision to own a car. Figure 4.6 suggests a preliminary summary of the link between car ownership, on the one hand, and household size and respondent's occupation, on the other hand. The data clearly show that there are more cars in larger households,

Figure 4.6. **Reported cars owned, by household size and respondent's occupation**

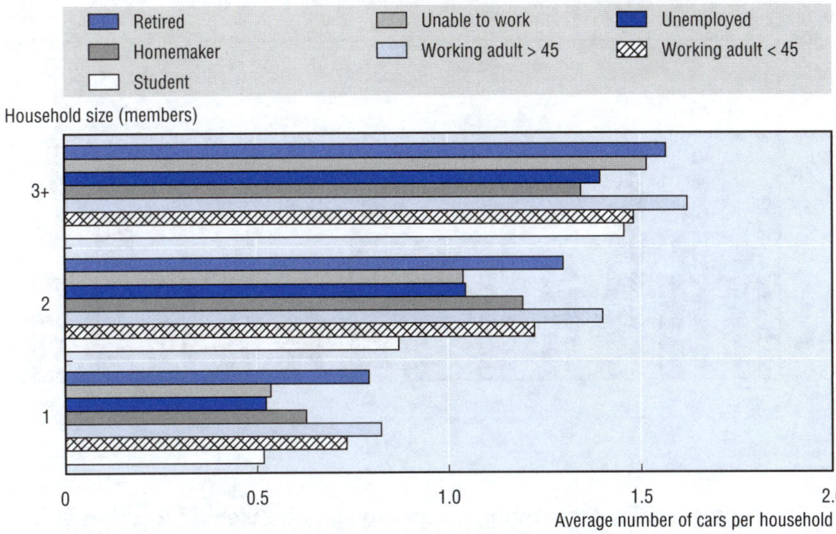

and in those of working adults. In single-person households, where car ownership is determined by the respondent alone, the latter pattern is the clearest. On the other hand, unemployed respondents' households have almost the same car ownership rate as those of working adults, when households are large. This is probably due to other persons in those households being employed themselves and owning cars.

Figure 4.7 shows that higher car ownership is robustly associated with higher household income. This holds for all eleven countries, with the expected fluctuations due to general income differences between the countries. Overall, households owning one car have about 50% more income than households owning none, while households owning two cars earn about twice as much as those with none.

The influence of environmental concern, as measured by the clustering procedure, appears to have a weak effect on car ownership. Although the level of concern is negatively correlated with car ownership (as intuition suggests), this correlation is mostly due to the indirect effect of household location. When accounting for this fact, the correlation between concern and car ownership is very weak, whereas the correlation between population density and car ownership is quite strong (as seen in Figure 4.3). This suggests that if environmental concern does strongly influence car ownership choices, then this effect is simultaneously determined with households' choice of location. This highlights an area in need of further empirical investigation.

Figure 4.7. **Reported number of cars owned per household, by income**

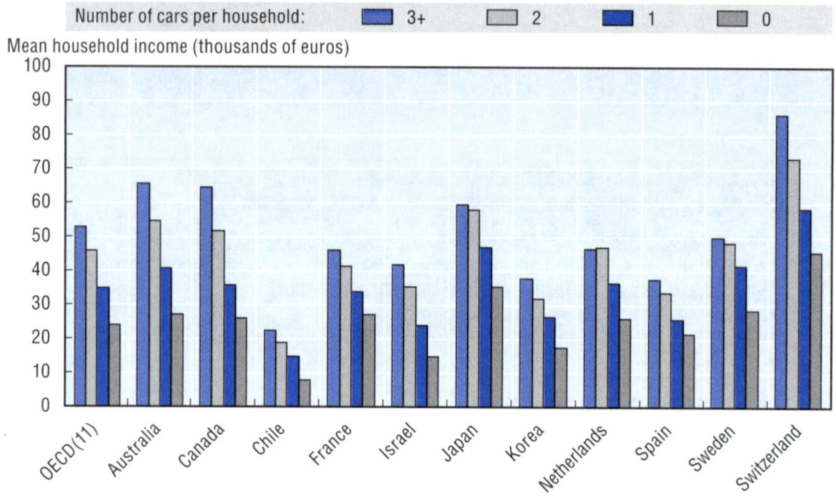

Figure 4.8. **Car ownership, air quality, and household location**

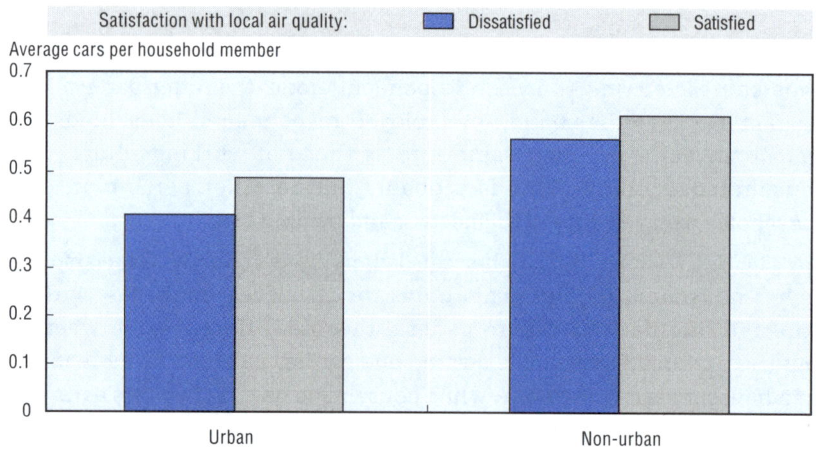

However, reported levels of satisfaction with local environmental quality appear strongly correlated with car ownership levels, even when controlling for related factors. Figure 4.8 shows average car ownership for households that express satisfaction or dissatisfaction with local air quality, whether the household is located in an urban or non-urban environment. Across different types of locations, respondents who are more satisfied with their local air quality own more cars on average than those who are dissatisfied with their air quality. Similar trends are exhibited for the other local environmental attributes covered in the survey. It is hard to determine whether households' opinions about their local air quality affect their decisions to purchase a car or vice versa: Households not owning cars may be more exposed to poor air quality (e.g. through more time spent outside) and thus more dissatisfied.

Determinants of decisions for not owning a car

Respondents in households not owning a car were asked why this was the case, by indicating the importance of various factors on a scale of 1 to 10. As can be seen in Figure 4.9, the most important reasons for not owning a car are that respondents cannot afford a vehicle or do not need one. Among respondents highly concerned with the environment, environmental concerns are the fourth most important reason for not owning a car, while they rank second-to-last for those in the low environmental concern cluster.

Figure 4.9. **Reasons for not owning a car, by environmental concern**

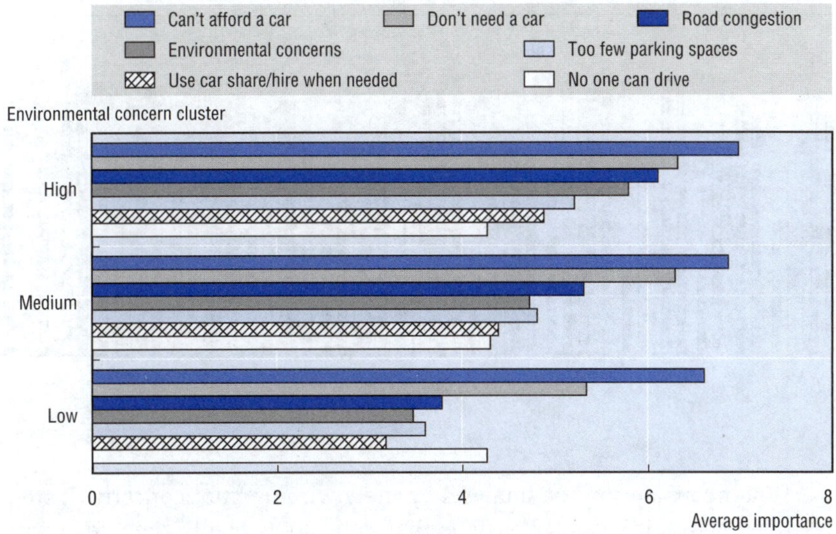

Interestingly, the importance of being able to drive and possessing a driving licence stays roughly the same across the different environmental concern clusters. Thus, no self-selection effect seems to be present at the level of driving licence ownership (that is, persons who are environmentally concerned are not less likely to possess a driving licence).

Factors influencing car choice

In addition to the decision whether or not to purchase a car (or cars), it is also of interest to consider how important environmental characteristics and fuel efficiency are relative to other vehicle attributes. While there is a rich body of research in the United States on the importance of vehicle attributes – including both fuel efficiency and polluting emissions – in purchase decisions, the focus here is on respondents' reported importance attributed to general factors in car purchase decisions. On a zero- to ten-point scale, respondents were asked how important eight different factors were in determining their car choice. As Figure 4.10 shows, safety appears to be the most important factor in the majority of countries, closely followed (and surpassed in Canada, Sweden, Australia and France) by both price and reliability. The least important factors are brand affinity and environmental impacts.

Figure 4.10. **Stated importance of factors in car purchases**

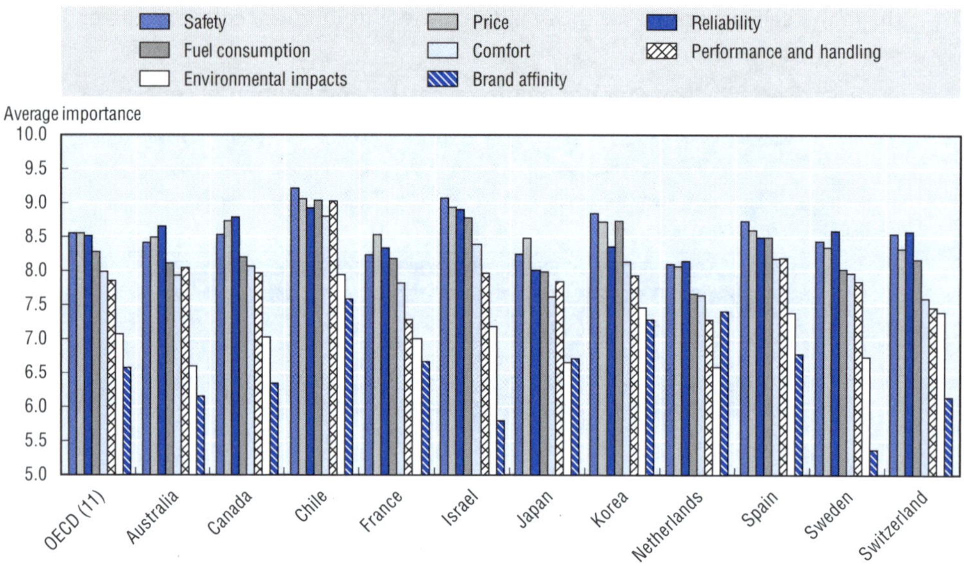

When considering the analysis by the environmental concern clusters described above, it becomes obvious that individuals highly concerned with environmental issues also place a higher emphasis on environmental impacts when choosing their cars. Interestingly though, the relative importance of environmental impacts, as compared to the other factors listed, remains quite low, regardless of the measured level of environmental concern (Figure 4.11).

Ownership of alternative-fuel vehicles and willingness to pay for electric cars

In addition to standard questions about car ownership, the 2011 EPIC Survey asked respondents, in particular, whether they owned any alternative-fuel vehicles (AFVs), as well as whether and how much they would be willing to pay for an electric car. AFVs here include hybrid vehicles,

Figure 4.11. **Importance of different factors in car purchases, by level of environmental concern**

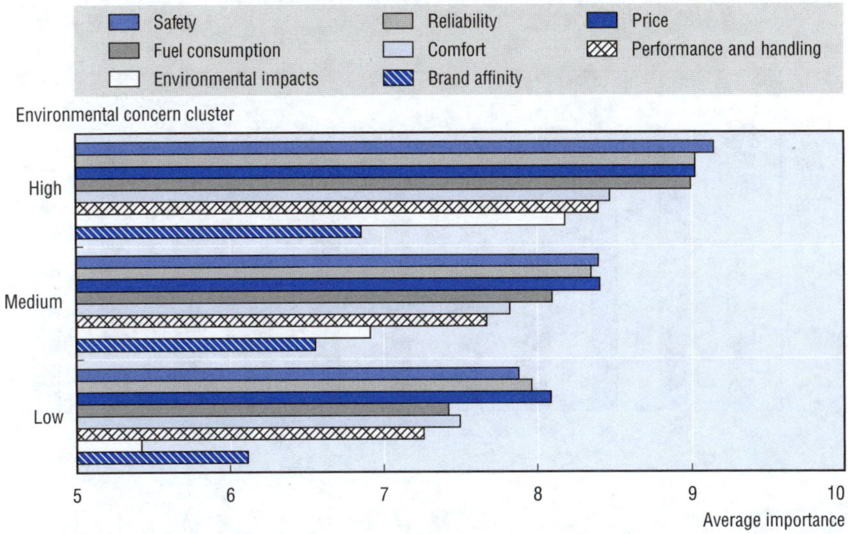

all-electric vehicles, as well as those fuelled at least in part by biodiesel, biogas, bioethanol or liquefied natural gas. All-electric cars, in particular, have received recent attention as an option for mitigating vehicle-related CO_2 emissions (when promoted in conjunction with renewable electricity generation). For example, Ireland has set a policy target of 10% electric cars in the vehicle fleet of the country by 2020 (Brady and O'Mahony, 2011).

Overall, stated ownership of AFVs across the surveyed countries ranges from less than 7% in Israel to 15% in Chile (Figure 4.12). The evidently high ownership rate of hybrid cars in Chile should be interpreted with caution and is likely to reflect the fact that some respondents misunderstood what constituted a hybrid car. Consistent with findings from other studies, it is clear that ownership rates of all-electric cars are very low across OECD countries.

It is of particular interest to know what characterises respondents who *a)* currently own AFVs and in particular hybrid electric vehicles, and *b)* would be willing to pay a price premium for an all-electric vehicle. In terms of socio-demographic attributes, only income and household size were found to be significantly associated with hybrid vehicle ownership. Surprisingly, among car owners, smaller, high-income households in the survey sample are less likely to own hybrids: A 1% increase in household income is associated with a 2% decrease in the probability of owning a hybrid (conditional on car ownership), whereas the presence of an additional member in a household is associated with a 1% increase in the likelihood of owning a hybrid , controlling for other factors such as education, population density around the household,

Figure 4.12. **Reported ownership of alternative-fuel vehicles by type and by country**

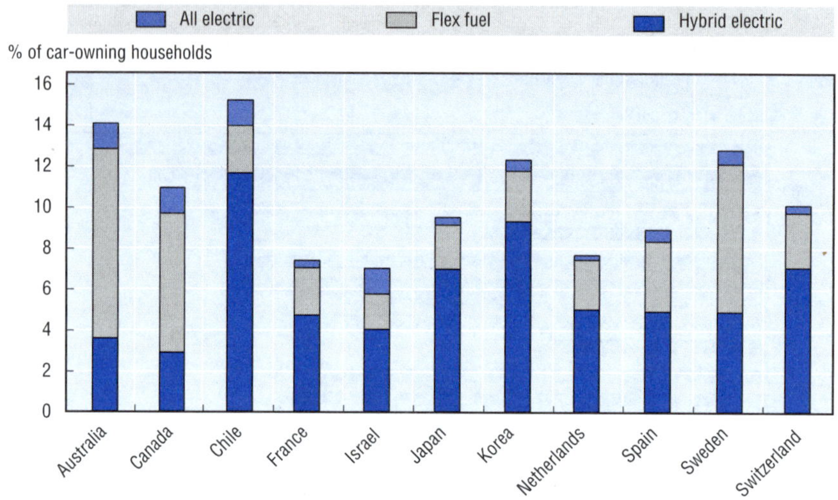

and countrywide idiosyncrasies. Given the very small percentage of hybrid owners and the multiple factors at play, further work is required to disentangle the distinct influence of different factors.

To measure the link between AFV ownership in general and specific attitudes about solutions to environmental problems, respondents were asked at the outset of the survey whether or not they agree with seven specific statements, shown in Figure 4.13 and discussed in more detail in Chapter 2. This figure shows that, of the seven statements, only opinions on whether future generations bear some responsibility for environmental problems seem to play a role in AFV ownership: the more respondents believe that future generations must deal with environmental problems, the more likely they are to own an AFV. This pattern demonstrates the subtle role that environmental attitudes play in behaviour. This particular statement may not only reflect respondents' sense of responsibility about environmental problems (in which case the above relationship is counter-intuitive), but may also reflect respondents' beliefs about the severity of future environmental problems.

Figure 4.14 shows the mean willingness-to-pay (WTP) for an electric car, by country. This additional WTP is given as the percentage increase in the price of a conventional car in exchange for having the car powered entirely through electricity. The average additional WTP over all respondents is about 20%. At its peak, it takes on values of as much as 38% for the Netherlands, 27% for Korea and 24% for Chile. The lowest WTP values are recorded in France (13%), Australia, Canada and Sweden (about 14% each).

Figure 4.13. **Environmental attitudes and ownership of alternative-fuel vehicles**

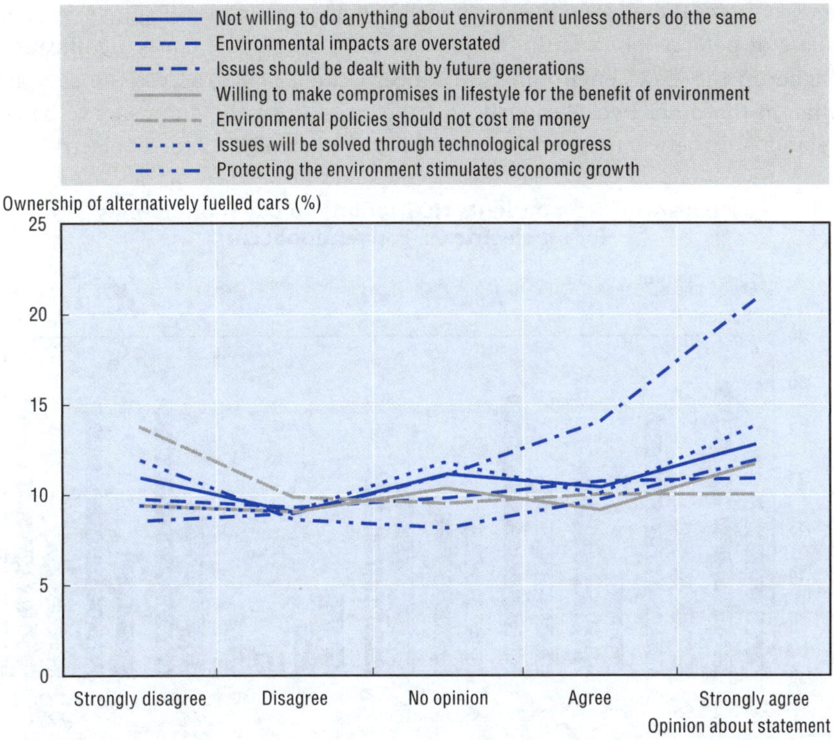

Figure 4.14. **Willingness-to-pay for an electric car versus a conventional car, by country**

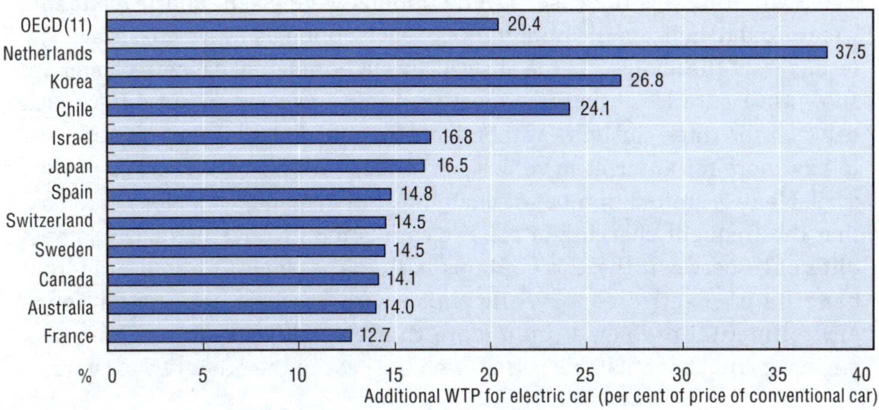

In total, 24% of the population would not be willing to pay more for an electric car relative to a conventional car. The relationship between WTP and environmental concerns is clearly visible across countries (Figure 4.15); the share of persons not wanting to pay more for an electric car is significantly higher in the "low" environmental concern cluster (38%, across the sample) than in the other two. Curiously, in Israel and Korea there appears to be no relationship between the environmental concerns and WTP for an electric car.

Figure 4.15. **Respondents NOT willing to pay more money for an electric vs. conventional car**

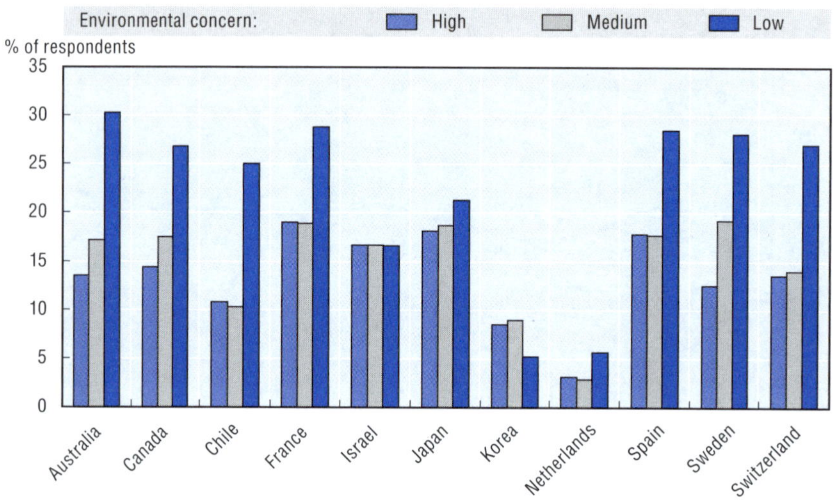

Respondents who stated a zero additional WTP for an electric car were asked why this was the case. Several options were given, and respondents were asked to indicate the importance of each option using a 1-to-4 scale. The results from the analysis of the answers by environmental concern cluster are shown in Figure 4.16. The three most prominent reasons do not differ much between the three clusters – respondents do not think that they should have to pay more for an electric vehicle, that adequate infrastructure is not yet available (e.g. limited number of charging stations) and, as a close corollary, that the frequent charging of electric cars is too inconvenient. The largest difference between the three groups is the assessment of environmental benefits of electric cars. While those who are less concerned about environmental problems often state that electric cars do not have any environmental benefits, this is less often the case in the other two clusters.

Figure 4.16. **Reasons for not wanting to pay extra for an electric car, by environmental concern cluster**

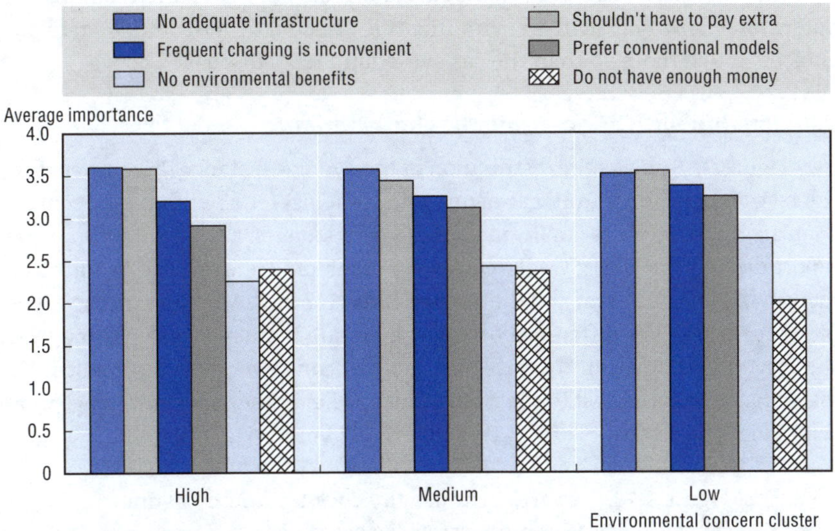

Another indication of how attitudes are reflected in respondents' decisions is found in how respondents' WTP for an electric car is linked to whether or not they find it acceptable to bear some of the cost of environmental policies. The hypothesis that people who accept that environmental policies will cost them extra money would have a higher WTP is confirmed by the data, as shown in Figure 4.17.

Figure 4.17. **Willingness-to-pay for electric car, by attitude towards costs of environmental policies**

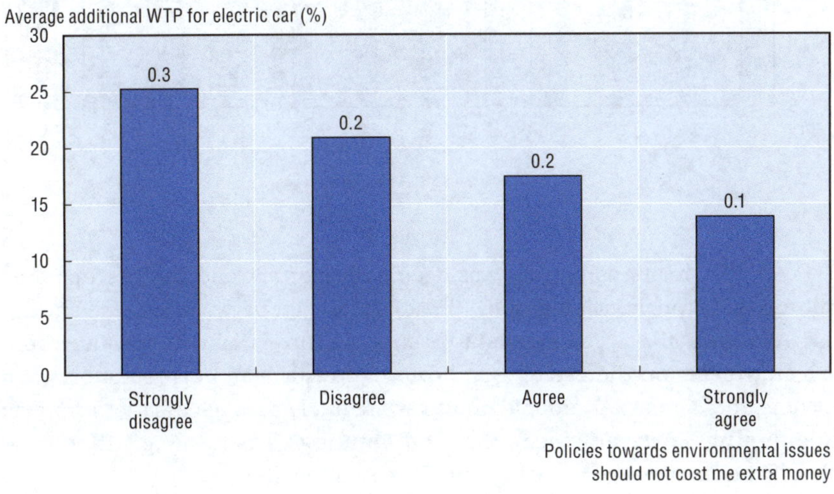

5. Car use

Car use, here expressed in distances driven weekly by the respondents, determines how heavily the environment is affected by individuals' transport choices, much more so than the car ownership rates discussed above.

Relationship with socio-economic characteristics

Figure 4.18 shows the distribution of the average distance driven weekly by the respondents (only from car-owning households) for all eleven participating countries, separately for male and female respondents. Car owners from Japan report driving the least (79 kilometres per week on average). The averages are highest in Israel (211 km) and the Netherlands (177 km). Another striking aspect of Figure 4.18 is the difference between men and women – males throughout the eleven participating countries drive significantly more than females. The discrepancy is the largest in the Netherlands, where men drive on average more than twice the distances driven by women (232 km vs. 113 km).

Figure 4.18. **Reported car use, by country and by gender**

Persons in car-owning households

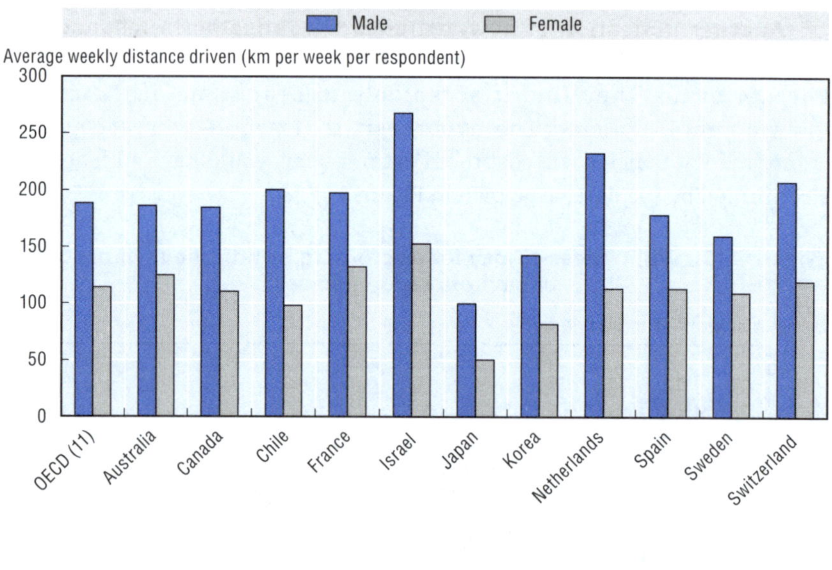

Another relationship that was deemed important and has been pointed out in several previous studies (OECD, 2011a) is that between income and car use. As shown above, car ownership is strongly correlated with income levels. When focusing on the use of cars, however, no correlation is found between income and car use. Additional, multivariate analysis is necessary to confirm this finding, however, since other confounding variables (e.g. urban/rural classification) may be obscuring a positive relationship.

Figure 4.19 shows the distribution of distances driven weekly by respondents from car-owning households as a function of the respondents' current employment status and household size. As can been seen, car use is highest among working adults of both age classes, and persons from large households tend to drive more. Unsurprisingly, home-makers and students cover the least number of kilometres on average.

Figure 4.19. **Reported car use, by employment status and household size**

Persons in car-owning households

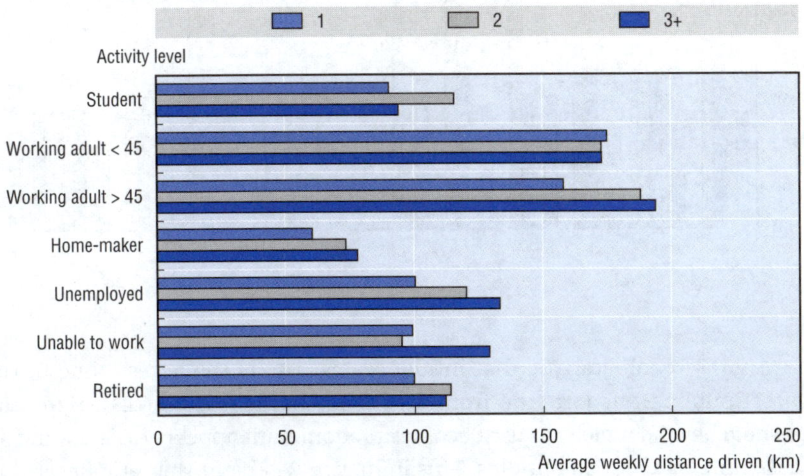

Note: "1", "2", "3+" on the right side of the figure represent the number of persons, one, two, three or more, in car-owning households

Relationship with spatial characteristics

Household car use, like car ownership, is heavily determined by residential location and the corresponding availability of alternative modes and destinations that are accessible by those modes. As Figure 4.20 shows, residents of major cities and suburbs who have good access to public transport are likely to drive less than people from rural areas: Car owners from urban areas drive less than 140 km on average during a typical week (i.e. less than 20 km a day). The difference in driving distances between urban and rural populations is the largest in Israel and Chile, where villages tend to be isolated and access to infrastructure such as jobs and shopping only possible by driving long distances.

It should be emphasised again that only respondents from households that possess at least one car are considered in this analysis, and thus the interaction with car ownership discussed in the previous section would amplify the relationships presented here. Furthermore, there is potential for a residential self-selection effect as mentioned before (households choosing to settle in urban areas at least in part because of better access to infrastructure such as public transport), which should be investigated in more detail.

Figure 4.20. **Reported car use, by country and by residential area type**

Persons in car-owning households

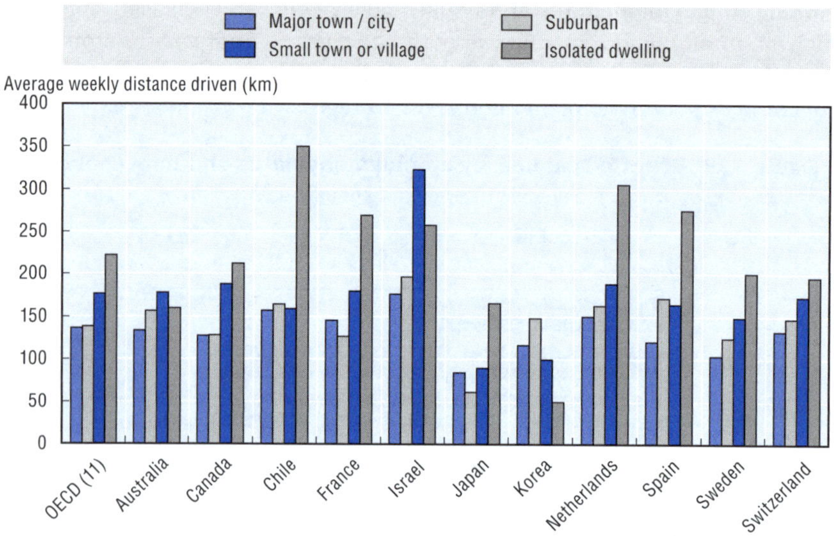

Another characteristic that influences car use is the access time to the nearest public transport stop from a person's home (Figure 4.21). Here, the threshold beyond which car use becomes predominant appears to lie around an access time of about 30 minutes. This influence will be further emphasised in the following section on household choice of transport mode for frequent trips.

Relationship with environmental concern and satisfaction with local environmental quality

It has been shown above that environmentally-concerned respondents are slightly less likely to own a car, although car ownership is high across all clusters. However, those respondents who stated that they were highly concerned with environmental issues and nevertheless own cars may act on their environmental attitudes by using their cars less, or only for trips where no alternative is available. Again taking the average distance driven weekly by car owners as an indicator, this assumption holds overall, though Korea and Chile are notable exceptions (Figure 4.22).

Whether there is a link between general environmental concern and car use, however, depends on which country is examined. Across countries, the difference in distance driven weekly between respondents from the "low" and "high" environmental concern clusters is only 10%, but this average hides heterogeneous patterns within countries. In seven countries, the expected relationship (more environmental concern translates into lower car use) is quite clear. However, in some countries such as Chile, Korea and Japan, the

Figure 4.21. **Reported car use, by access time to nearest public transport stop**

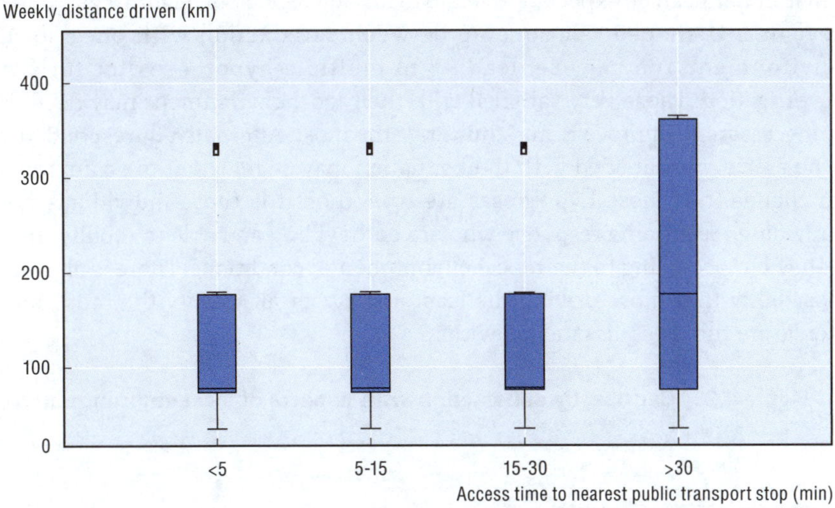

Note: Shaded bars represent 75% of the data, with the black lines inside indicating the mean. Brackets represent 95% of the data, and the black circles near the top are outlying observations.

Figure 4.22. **Car use, by country and by environmental concern cluster**

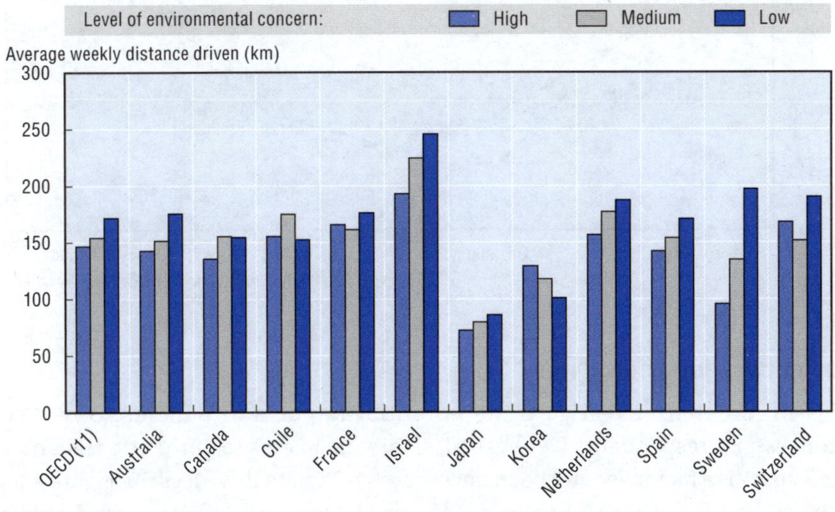

relationship is inverted. These data suggest that environmental issues may only have a marginal impact on respondents' driving behaviour in those countries, and to confirm this relationship it would have to be measured in conjunction with other possible covariates. There may also be a need to explore other ways of measuring environmental attitudes in an analysis of these data.

As with car ownership, perceptions and attitudes towards one's local environment can be expected to relate to driving behaviour. Figure 4.23 clearly reveals a U-shaped relationship between satisfaction with one's local environment and car use, leading to multiple hypotheses for further investigation: Those very satisfied with their local environment may deem it unnecessary to improve it, and thus drive the most. Alternatively, respondents who are very unsatisfied with their situation may be reluctant to do anything to change it. If these hypotheses are true, then it is those individuals not providing an extreme response who are both willing and able to modify their behaviour according to their local environmental conditions. There is also the possibility that those driving the least *notice* poor air quality the most, and hence are the most dissatisfied with it.

Figure 4.23. **Car use, by satisfaction with aspects of local environment**

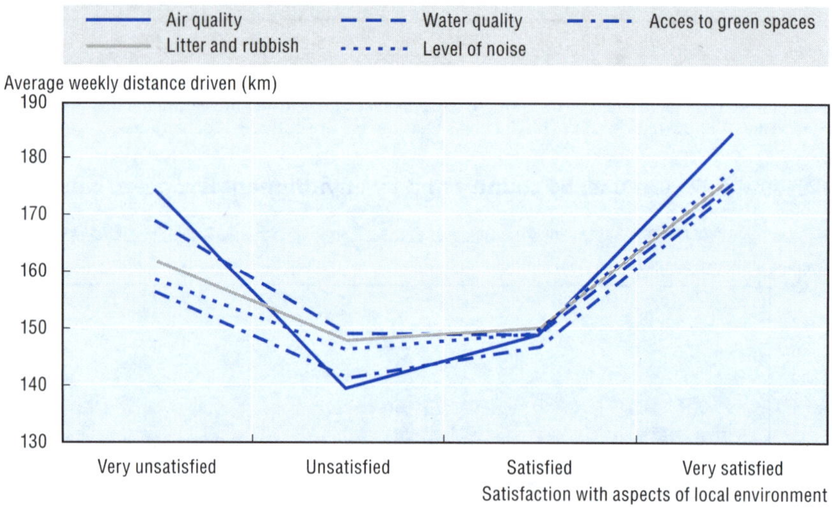

Incentives for driving less

In order to investigate the potential impacts of policies to reduce congestion, respondents were asked to rate on a zero- to ten-point scale how effective five factors would be in encouraging them to drive less. As Figure 4.24 shows, the factor viewed as most effective in this regard is improved public transport options. In the same vein, the addition of more and safer cycling paths would also encourage the use of alternative modes, particularly in Chile, Korea and Japan. Surprisingly, the expected impact of increasing the costs for car use (e.g. through higher fuel taxes) appears quite moderate. This could reflect a strategic bias in responses to this question, i.e. respondents did not want to encourage the implementation of higher fuel taxes.

Figure 4.24. **Reported assessment of measures to reduce car use**

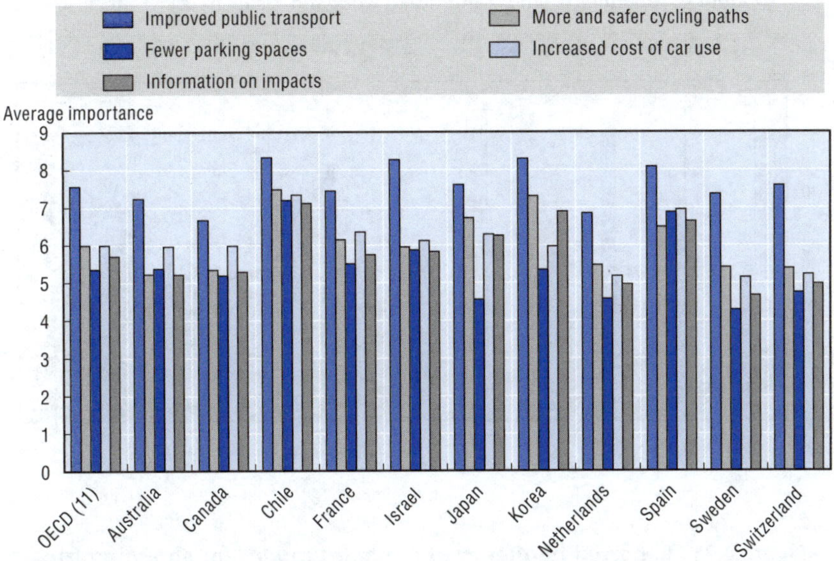

Parking availability is the policy tool which gains the least support. This is surprising, at least for Switzerland. A recent study conducted at the Institute for Transport Planning and Systems (IVT) by Axhausen et al. (2011) emphasised parking policies, and significant effects on mode choice were found. As with higher costs of car use, it is possible that responses as to the importance of this factor suffer from the same strategic bias, i.e. respondents do not want to encourage a reduced supply of parking spaces. Indeed, some respondents may be even more averse to policies which constrain their consumption choices.

6. Household choice of transport mode for frequent trips

In addition to car ownership and use, respondents also reported how they travelled to work and their most common shopping location. They were asked to specify their most common mode for reaching those destinations, the duration of those trips, and the attributes of the available alternatives. As Figures 4.25 and 4.26 show, the most common modes for travelling to work for urban residents is either by car (Australia, Canada, France, Israel, Spain) or by public transport (Chile, Korea, Switzerland). In Sweden and Japan the two modes are approximately equally common. In the Netherlands and Sweden cycling to work is common. Walking to work is prevalent in Spain, France, Japan and Switzerland.

Figure 4.25. **Reported frequency of transport modes for travelling to work**

Among respondents who commute to work and reside in urban areas

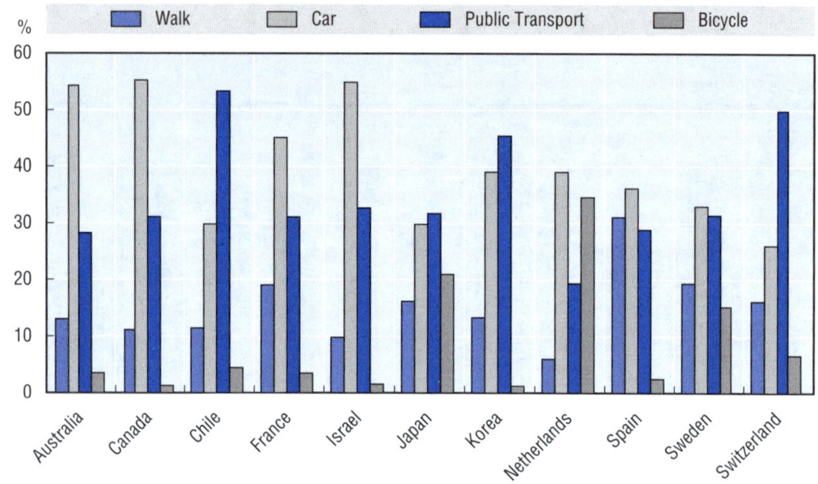

Figure 4.26. **Reported frequency of transport modes for shopping trips**

Among applicable respondents

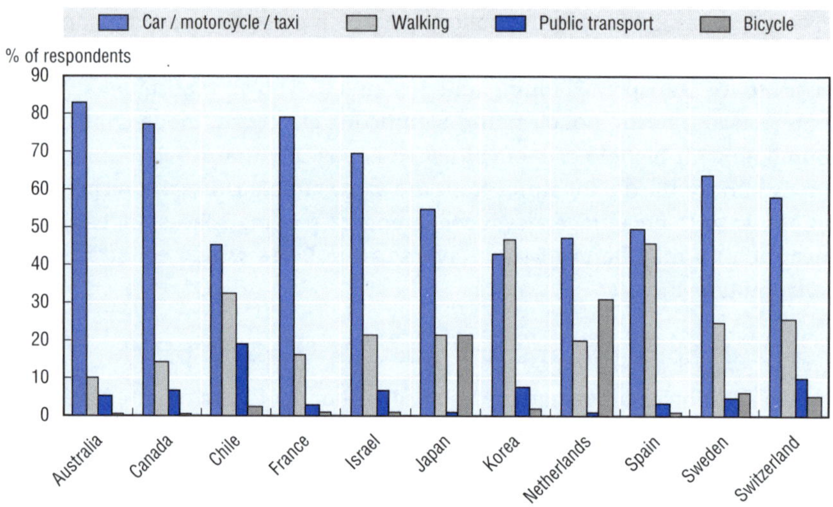

The most common transport mode for shopping is by private motor vehicle, followed by public transport. Exceptions to this trend can be seen in Chile and Japan, where public transport is more heavily used for commuting than other modes, and in Korea where respondents reported walking more often than driving to their most common shopping area. The share of walking to work is largest in Spain, while the bicycle has the largest share in the traditionally bike-affined Netherlands (Dijkstra and Pucher, 2003).

In comparing Figures 4.25 and 4.26, one observes that public transport plays practically no role in shopping trips in any of the eleven countries: a large majority of trips are carried out either by car or on foot. This is in line with recent Swiss results (Axhausen et al., 2011), where driving was shown to be inherently preferred to other modes for shopping trips. In theory, choice of transport mode is influenced by the distance and travel time to the destination, as well as by the quantity and nature (i.e. bulk and weight) of the purchased goods. Australia and Canada again have the largest shares of car trips, which may be explained by shopping opportunities often being located in malls on the outskirts, as opposed to small, nearby shops.

As a general trend, Figure 4.27 shows the influence that household income has on mode choices. It can be seen that for both commuting and shopping trips, respondents travelling by car tend to have higher incomes than those typically using public transport. The following subsections focus on factors related to commuting and shopping trips, as well as on potential incentives to use public transport.

Figure 4.27. **Mean annual income by type of trip and most frequent mode**

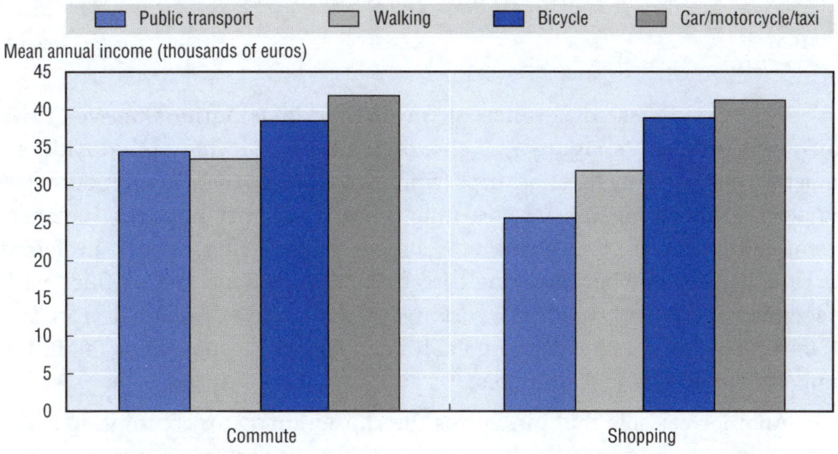

Access to alternative modes of transport

As Figure 4.28 shows, the choice of a commuting mode is heavily dependent on the corresponding travel time. When a personal car is the main mode for commuting (see bottom horizontal bar), it is the shortest available alternative three-quarters of the time (that is, the travel time for the fastest alternative is at least 5 minutes longer, and in the majority of cases over 30 minutes longer). On the other hand, respondents travelling by other modes often do so despite there being faster available alternatives (mostly private cars). This is a first indicator that people trade off travel time against other factors (financial, environmental, stress) when considering commuting by car or other means.

Figure 4.28. **Reported commuting modes, by travel time with fastest alternative**

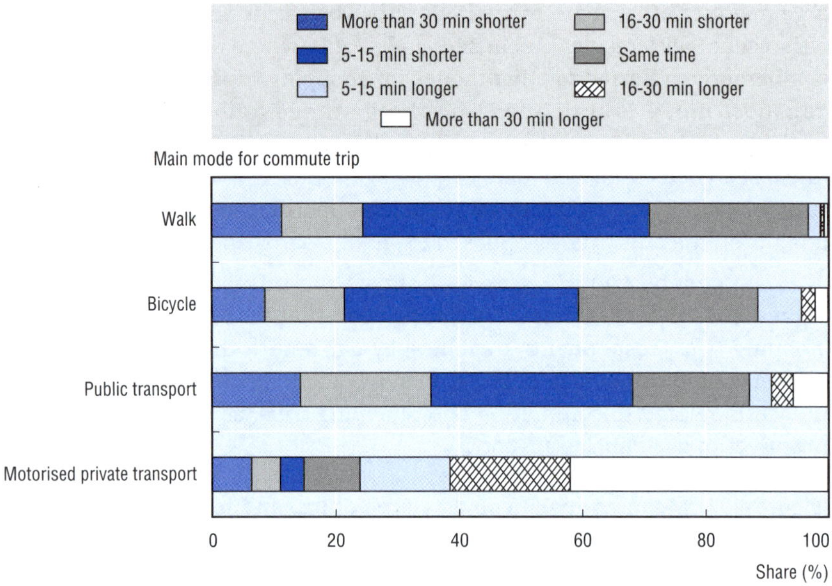

There is a clear difference of travel time distributions between the different modes, as Figure 4.29 shows. Walk and bicycle are considered practical options only for short trips (where it takes less than 30 minutes to get to work by those means). On the other hand, a majority of public transport commuting trips are over 45 minutes long. Here, public transport is preferred to driving, probably because long trips by public transport can be filled with secondary activities (reading, working, etc.; see for example Ohmori and Harata, 2008, for a corresponding analysis in Japan). Most car trips fall in the range between 15 and 45 minutes.

Another variable that influences the chosen mode for commuting is the access time to the nearest public transport stop. As Figure 4.30 shows, public transport is the most likely to be chosen when the access time to the stop is short. When the nearest stop is more than 15 minutes away, the share of public transport drops off considerably, in favour of higher shares for the car alternative.

For shopping trips, the nature of the destination (e.g. convenience store, or large supermarket) was found to be significantly tied to the choice of transport mode. Respondents were asked where they shop for food in a normal week, with several possible answers being offered. Figure 4.31 shows the shares of main shopping modes for each of the options. Expectedly, the share of car trips is highest for respondents who frequently shop at out-of-town,

Figure 4.29. **Distribution of commuting travel times, by mode**

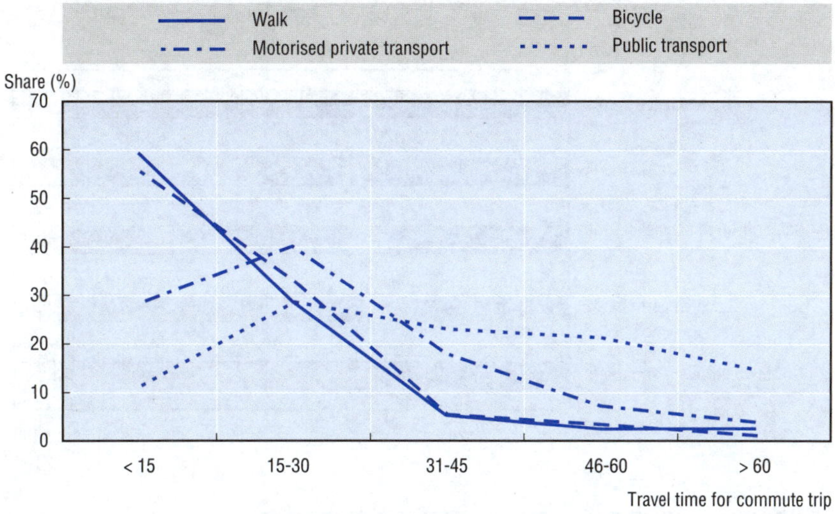

Figure 4.30. **Main mode for commuting and access time to public transport stop**

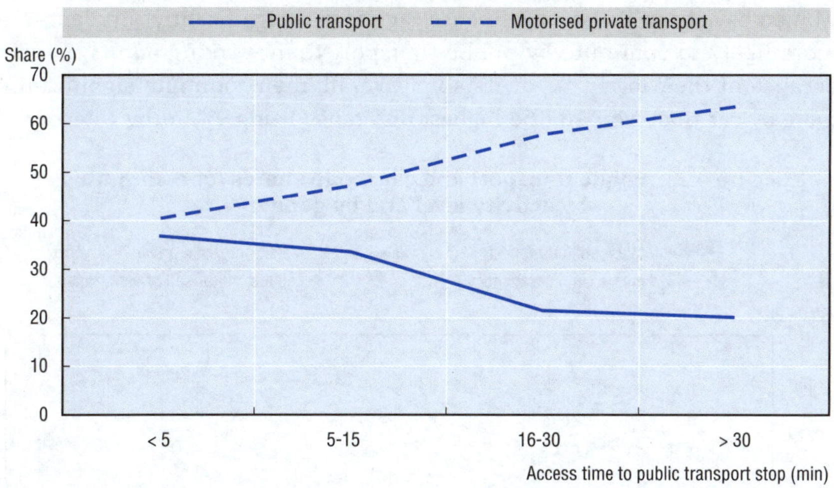

large supermarkets. For local supermarkets, specialist shops, markets and convenience stores, walking becomes a more dominant mode, as those stores are probably less distant from respondents' home locations. Overall, though, the share of car trips is high, while those for public transport and bicycles are negligible.

Figure 4.31. **Main mode for shopping trips, by store type**

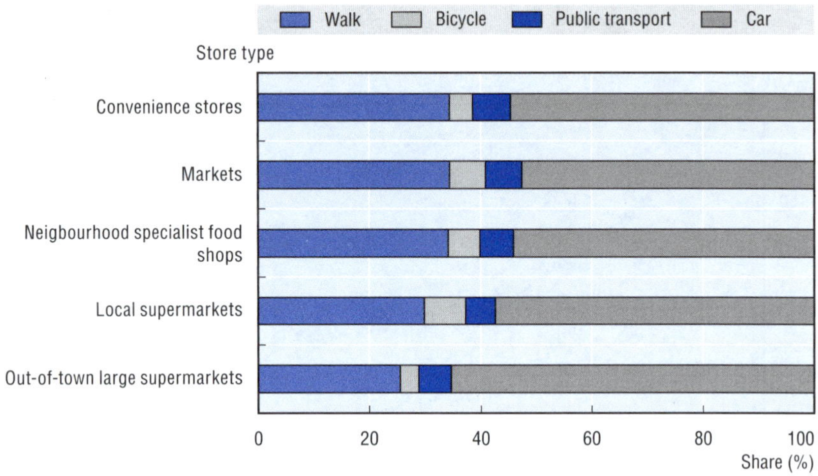

Relationship with socio-economic characteristics

Respondents' commuting modes are strongly associated with their socio-economic characteristics. The influence of the respondents' activity level in interaction with their gender is shown in Figure 4.32 (as men and women have been shown to influence car use very significantly). Students are more likely to commute by public transport than working adults, largely because of their lower car ownership. Overall, men commute significantly more by car than women (the highest difference being 9% in the category of

Figure 4.32. **Public transport and car mode shares for commuting, by activity level and by gender**

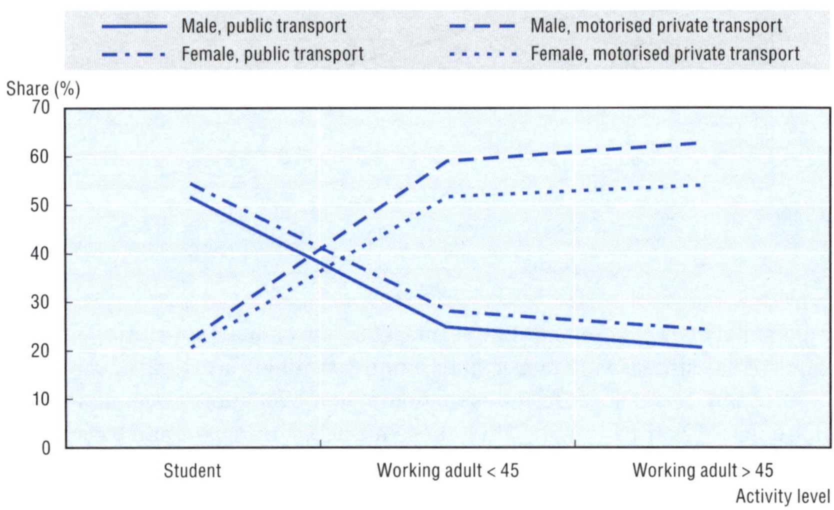

working adults above the age of 45). This is consistent with the above finding that men travel much longer distances than women in general, and is in line with recent results for Switzerland (Axhausen and Weis, 2009). For shopping trips, similar relationships (not shown) between mode choice, on the one hand, and gender and activity level, on the other, were observed.

Relationship with household location

As is the case for car use, respondents' residential location evidently is closely related to their choice of transport mode, as shown in Figure 4.33. While the shares of public transport for commuting trips are high for respondents living in cities or suburban areas, they are significantly lower in rural areas. There are, however, huge differences in these trends among the eleven participating countries. Especially in Chile and Switzerland, the share of public transport among residents of isolated dwellings is very high.

Figure 4.33. **Public transport mode shares for commuting, by country and by residential area type**

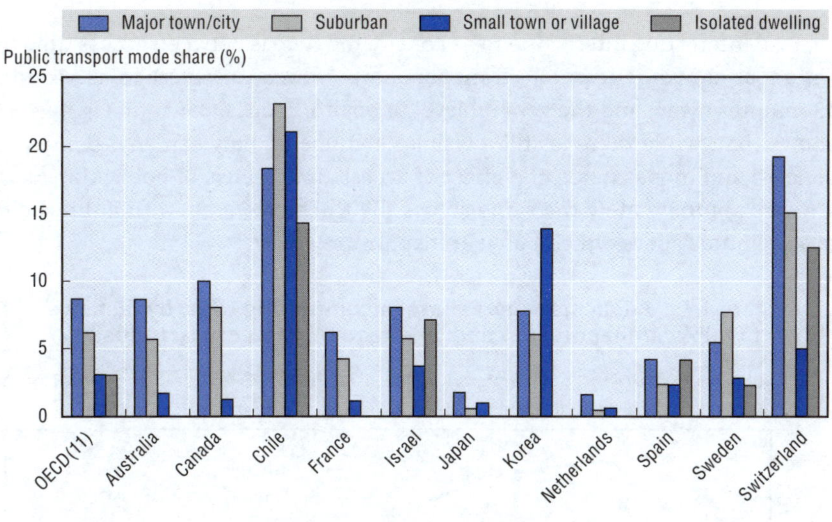

Relationship with environmental concern

The relationship between environmental attitudes on mode choice is shown in Figure 4.34. Here, it can be seen that the influence of environmental concern is quite large. Respondents highly concerned with environmental issues commute by car 15% less than those least concerned with these issues. Interestingly, increasing environmental awareness appears only to involve a shift from car use to public transport; walking and cycling rates are roughly constant across different levels of environmental awareness.

Figure 4.34. **Main mode for commuting, by environmental concern cluster**

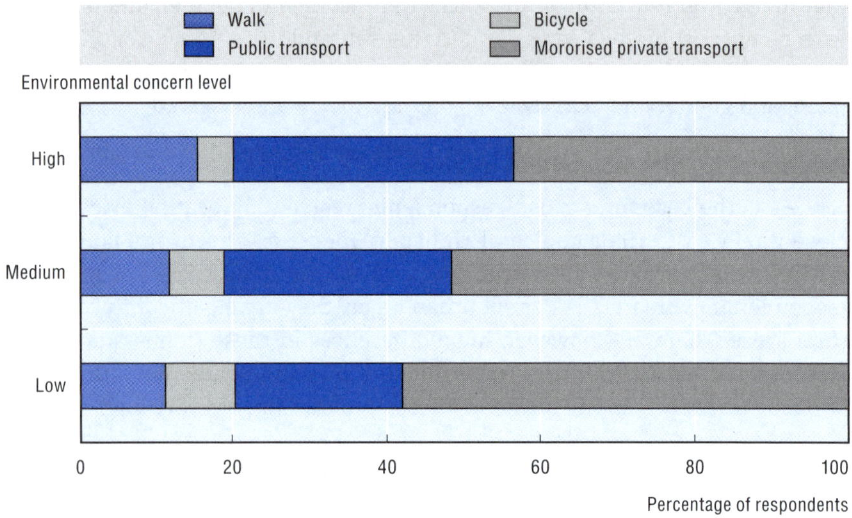

An interesting interaction to be considered here is between access time to the most convenient public transport stop (which, on average, is around 15 minutes away) and the general level of environmental concern. Figure 4.35 shows that, while there is still a visible relationship between environmental concern and mode choice, the effect of access time on use of public transport appears approximately the same across the three clusters – with a threshold of 15 minutes above which public transport use is low.

Figure 4.35. **Public transport share in commuting trips, by distance to public transport stop and by environmental concern cluster**

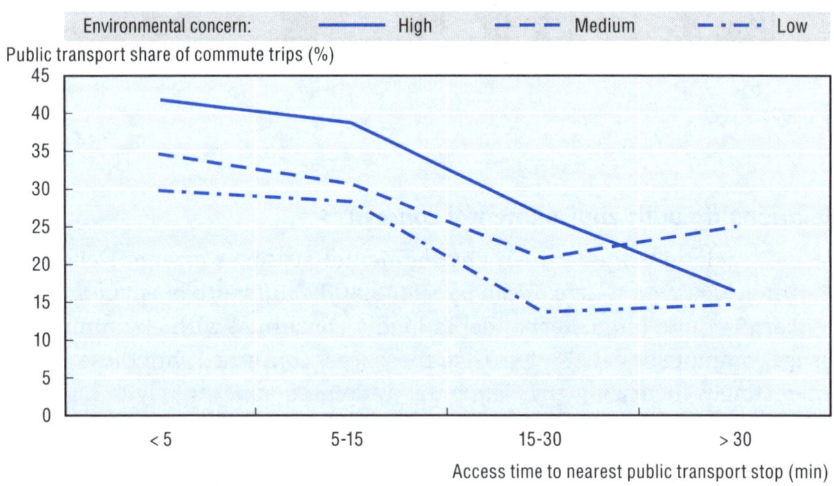

Incentives to use public transport more

When attempting to encourage travellers to use their cars less and to switch to alternative modes, it is important to know which policies could be applied to encourage such changes. Figure 4.36 plots the importance that respondents attributed to the various types of improvements in public transport, by respondents' level of environmental concern.

Figure 4.36. **Factors encouraging public transport use, by environmental concern cluster**

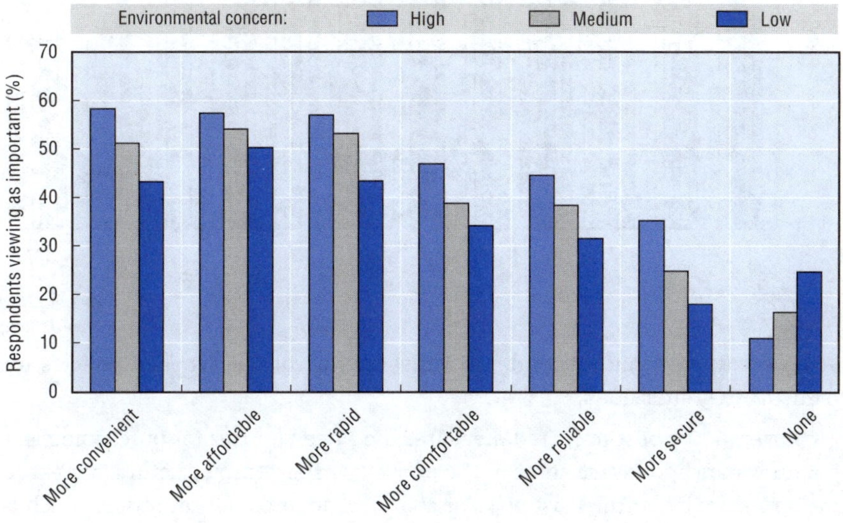

The most important issues overall are convenience (probably related to the access time described above), cost and speed: over half of those surveyed indicated these as important for promoting public transport use. Of secondary importance are comfort and reliability, while security is surprisingly the least supported issue. In total, only 17% of the respondents said that none of the incentives would lead them to travelling more by public transport; thus, there seems to be an overall willingness to switch transport mode, provided that the appropriate policies to deal with the main issues are put in place. Overall, the data suggest that respondents more concerned in general with environmental issues would more quickly respond to any of the incentives than the others.

7. Support for government policies to reduce vehicle CO_2 emissions

When thinking of what measures to implement to reduce CO_2 emissions, authorities will be interested in how well different measures are accepted by the population. To that end, respondents were asked how strongly they would support a list of actions, using a zero- to ten-point scale. Figure 4.37 shows the

Figure 4.37. **Support of government actions to reduce CO_2 emissions, by country**

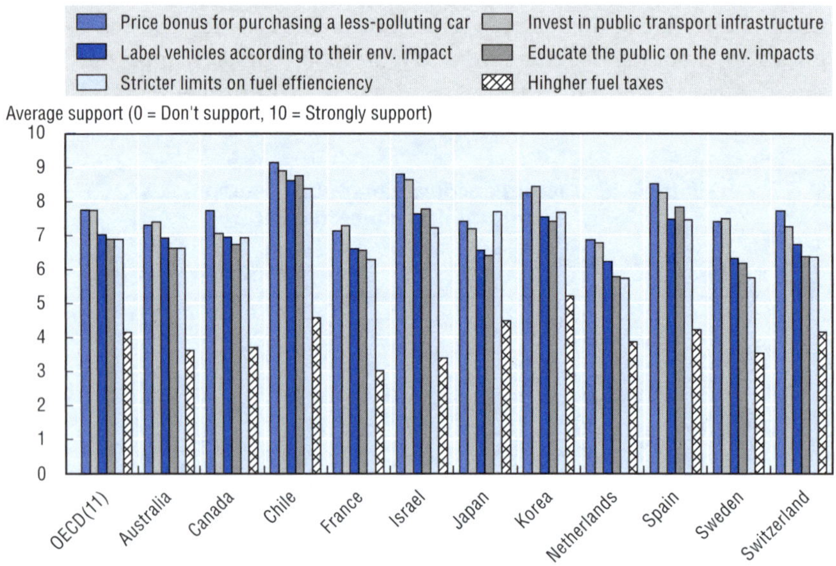

levels of support that respondents indicated for various types of policies for reducing CO_2 emissions.

Overall, the best-accepted measures are price bonuses for purchasing less polluting cars and investments in the public transport infrastructure. Increased fuel taxes are by far the least popular course of action in all countries, which is consistent with the low enthusiasm found for policies which increased the cost of car use (Figure 4.24). A notable exception is Japan, where imposing strict limits on vehicle fuel efficiency is the best-rated measure, whereas it is a very unpopular one in other countries. The broad factors behind these attitudes towards policies are discussed in more detail, and in relation to other environmental policy domains, in Chapters 1 and 2. Across countries, there seems to be no large differences in the relative popularity of these measures.

8. Conclusions

The analysis presented here is aimed at providing a broad overview of the patterns that can be found in the OECD 2011 EPIC data relating to the transport sector. A more thorough analysis is necessary to investigate whether the associations highlighted above are causal: for example, does the significant correlation between car ownership and satisfaction with local air quality mean that car ownership and use affect perceptions of local air quality? Or are those who are unsatisfied with their local air quality less likely to purchase a car? Econometric analysis is necessary to address these questions.

Nevertheless, the descriptive analysis presented here suggests some important lessons for transport policies targeted on households. Statistically, socio-economic and spatial characteristics are strong predictors of transportation decisions and preferences. Environmental attitudes – as measured by the clustering procedure described at the beginning of this chapter – are weaker statistical predictors of transportation decisions, except in some specific cases mentioned below. Table 4.1 presents a summary of the main effects on car ownership and use. In brief, availability, affordability and speed of public transport were identified as the major drivers of a potential change in behaviour.

Table 4.1. **Main effects of socio-economic and attitudinal variables on car ownership and use**

Category	Level	Number of cars		Weekly distance
		Per household	Per adult	Driven [km]
-	Sample average	1.31	0.58	154.2
Country	Australia	1.66	0.78	155.8
	Canada	1.39	0.68	148.2
	Chile	0.94	0.35	160.9
	France	1.59	0.75	165.7
	Israel	1.24	0.53	210.6
	Japan	1.24	0.51	78.6
	Korea	1.17	0.44	121.8
	Netherlands	1.23	0.59	177.0
	Spain	1.46	0.60	150.3
	Sweden	1.21	0.66	138.1
	Switzerland	1.27	0.62	162.8
Gender	Male	1.32	0.58	187.7
	Female	1.29	0.58	113.9
Activity level	Student	1.24	0.44	96.8
	Working adult < 45	1.32	0.62	172.5
	Working adult > 45	1.43	0.61	186.6
	Home-maker	1.29	0.56	76.4
	Unemployed	1.12	0.51	124.8
	Unable to work	1.04	0.53	107.4
	Retired	1.25	0.62	111.3
Household size	1	0.71	0.71	142.6
	2	1.25	0.64	152.0
	3+	1.49	0.55	157.8
Type of residential location	Major town/city	1.10	0.48	136.6
	Suburban	1.38	0.61	138.2
	Small town or village	1.50	0.69	176.3
	Isolated dwelling	1.81	0.88	222.1
Environmental concern cluster	high	1.24	0.53	146.3
	medium	1.34	0.60	153.7
	low	1.39	0.64	170.9

It is unsurprising, but encouraging, to note that investments in public transport infrastructure are very popular (the willingness to pay for public transport services is not estimated, however, and is perhaps a topic for future rounds of the survey). Where such investments are possible, they could lead to further shifts away from more polluting transport modes. In the same vein, premiums for buying less polluting cars would help as well. The underlying willingness to invest in such cars is present, as evidenced by the data on the willingness to pay additional costs for electric cars. However, the lack of infrastructure for using these vehicles still limits their adoption.

Notably, environmental attitudes are closely tied to choice of car and willingness to pay for environment-friendly vehicles. Attitudes are also associated with the choice of transport mode: more environmentally "aware" respondents use public transport more often. However, the issue of causality deserves particular scrutiny in this case: Do more environmentally aware individuals use public transport more, or does the use of public transport promote environmental awareness? The likely answer is a combination of the two. Effective policies for increasing the use of existing public transport systems could potentially include promoting environmental awareness among the general population and also decreasing the cost or inconvenience of public transport among environmentally aware groups (perhaps by including coupons for the use of public transport services with the purchase of water- and energy-efficient appliances).

References

Axhausen, K.W. et al. (2010), "Models of mode choice and mobility tool ownership beyond 2008 fuel prices", *Transportation Research Record*, pp. 2157.

Axhausen, K.W. et al. (2011), *Influence of parking on location and mode choice: A stated choice survey*, Paper presented at the 91st Annual Meeting of the Transportation Research Board. Washington, DC.

Axhausen, K.W. and A. Simma (2004), "Interactions between travel behaviour, accessibility and personal characteristics: The case of the Upper Austria region". *European Journal of Transport and Infrastructure Research*, Vol. 3, pp. 179-198.

Axhausen, K.W. and C. Weis (2009), "Induced travel demand: Evidence from a pseudo panel data-based structural equations model", *Research in Transportation Economics*, Vol. 25, pp. 8-18.

Brady, J. and M. O'Mahony (2011), "Travel to work in Dublin. The potential impacts of electric vehicles on climate change and urban air quality", *Transportation Research Part D: Transport and Environment*, Vol. 16, pp. 188-193.

Beevers, S.D. and D.C. Carslaw (2002), "The efficacy of low emission zones in central London as a means of reducing nitrogen dioxide concentrations", *Transportation Research Part D: Transport and Environment*, Vol. 7, pp. 49-64.

Bocarejo, J.P. and R. Prud'Homme (2005), "The London congestion charge: a tentative economic appraisal", *Transport Policy*, Vol. 12, pp. 279-287.

Cao, Xinyu, Mokhtarian, P. L., and Handy, S. L. (2009), "Examining the impacts of residential self-selection on travel behaviour: a focus on empirical findings", *Transport Reviews*, Vol. 29, No. 3, pp. 359-395.

Chanel, O. et al. (2000); "Public-health impact of outdoor and traffic-related air pollution: a European assessment", *The Lancet*, Vol. 356, pp. 795-801.

Dargay, J. and M. Hanly (2004), *Land use and mobility*, 10th World Conference on Transport Research. Istanbul.

Dargay, J. and G. Giulliano (2006), "Car ownership, travel and land use: A comparison of the US and Great Britain", *Transportation Research Part A: Policy and Practice*, Vol. 40, pp. 106-124.

Dijkstra, L. and J. Pucher (2003), "Promoting safe walking and cycling to improve public health: Lessons from the Netherlands and Germany", *Public Health Matters*, Vol. 93, pp. 1509-1516.

Dill, J., S. Handy and J. Pucher (2010), "Infrastructure, programs, and policies to increase bicycling: An international review", *Preventive Medicine*, Vol. 50, Supplement, pp. S106-S125.

Jonas, E. (2009), "A cost-benefit analysis of the Stockholm congestion charging system", *Transportation Research Part A: Policy and Practice*, Vol. 43, pp. 468-480.

Meltzer, J. (2012), "Climate Change and Trade – The EU Aviation Directive and the WTO", *Journal of International Economic Law*.

Mian, A. and A. Sufi (2010), "The Effects of Fiscal Stimulus: Evidence from the 2009 'Cash for Clunkers' Program", *National Bureau of Economic Research Working Paper Series*, No. 16351.

Mokhtarian, P.L. (2008), "Examining the impacts of residential self-selection on travel behaviour: A focus on empirical findings", *Transportation Research B*, Vol. 42, No. 3, pp. 204-228.

Norusis, M. (2011), *IBM SPSS Statistics 19 Advanced Statistical Procedures Companion*, Upper Saddle River, Prentice Hall.

OECD (2010), *Transport Greenhouse Gas Emissions: Country Data*, International Transport Forum (ed.), OECD Publishing.

OECD (2011), *Greening Household Behaviour: The Role of Public Policy*, OECD Publishing. doi: *10.1787/9789264096875-en*.

OECD (2011b), OECD.Stat (database). doi: *10.1787/data-00285-en*.

Ohmori, N. and N. Harata (2008), "How different are activities while commuting by train? A case in Tokyo", *Tijdschrift voor Economische en Sociale Geografie*, Vol. 99, No. 5, pp. 547-561.

World Bank (2008), *Motor vehicles per 1000 people* [Online]. Washington, DC. Available: *http://data.worldbank.org/indicator/IS.VEH.NVEH.P3* [Accessed 31 October 2011].

Greening Household Behaviour
Overview from the 2011 Survey

Chapter 5

Household behaviour and water use

by
Quentin Grafton*

This chapter presents an overview of the survey data on the determinants of households' water use and looks at the impacts of policy measures such as water pricing and appliance labelling. It examines the determinants of water-saving behaviour and investment in water-saving appliances and whether having to pay for water according to volume actually reduces consumption. The role of respondents' environmental "norms" is also analysed.

* The Australian National University.

The statistical data for Israel are supplied by and under the responsibility of the relevant Israeli authorities. The use of such data by the OECD is without prejudice to the status of the Golan Heights, East Jerusalem and Israeli settlements in the West Bank under the terms of international law.

1. Introduction

Water plays a crucial role in the development, growth and sustainability of local communities. In recent years, water scarcity has become a global environmental problem. Growing populations and higher per-capita demand for water, together with more volatile supplies, have made water management an increasingly important issue for water authorities worldwide.[1]

The traditional approach to water management has been to focus on supply-side policies. However, with the rise in infrastructure costs due in part to the increasing stringency of environmental regulations and increased water scarcity, expansion of supply requires increasingly large capital investments. This makes supply-side policies more difficult to implement, especially in a context of significant public funding constraints. Moreover, supply expansion also requires several years of planning before the water becomes available.

In response to this, there have been changes in the approach to water management. In particular, "integrated resource planning" is increasingly common, with analyses of capacity expansion complemented with alternative strategies for demand-side management (Terrebonne, 2005). Further, because of the effects of climate change, water supply in many areas is becoming more and more variable, resulting in imbalances between supply and demand (Arbués et al., 2003). Thus, demand-side policies have become increasingly important and many countries have placed a strong emphasis on demand-side management.

This chapter looks at demand-side policies for water management, focusing on household water consumption. Using data collected from more than 12 000 households across eleven OECD countries, it analyses factors that affect households' water-saving behaviours and adoption of water-efficient devices. Also studied are the factors driving household satisfaction with the quality of tap water. It examines the correlation between satisfaction and the primary sources of drinking water across households. The study is primarily descriptive.

The study examines the effect of both volumetric water charges and attitudinal characteristics such as environmental concerns, perceptions and attitudes, on undertaking water-saving behaviours and investing in water-efficient devices, as well as reported satisfaction with tap water quality. Given that little attention has been paid to the influence of attitudes, perceptions and values on household environmental behaviours (Van Den Bergh, 2008), the findings of this research are informative and provide important policy implications for water authorities and policy makers who wish to encourage water-saving behaviours, to induce adoption of water-efficient equipment, and to encourage people to drink high-quality water straight from the tap instead of drinking purified, boiled or bottled water.

This chapter first provides a brief review of research on water conservation. It then describes the dataset used and presents the empirical results. Policy insights arising from the results are discussed before a presentation of the conclusions. An appendix gives useful information on the definition of variables.

A summary of key findings on residential water use is provided in Box 5.1.

Box 5.1. **Residential water use: key findings**

Findings from descriptive analysis suggest that:

- Unit-pricing for water, through the use of a volumetric water charge, unambiguously increases households' efforts at water conservation. This includes both water-saving behaviours and financial investments in water conservation.
- Low-income households more frequently engage in water-saving behaviours, but are somewhat less likely to invest in water efficiency improvements. Need-based grants for water efficiency investments could provide an important means of improving water conservation by targeting this income group.
- Households who rent rather than own their homes engage less often in water-saving behaviours and make fewer financial investments in water efficiency. Programmes for increasing awareness and promoting water-saving investments among tenants could be a useful way to correct this economic distortion.
- Households' concern for various environmental issues and level of community involvement is significantly correlated with adoption of water-efficient devices. A higher level of concern about natural resource depletion and environmental issues, supporting or participating in an environmental organisation, having voted in the past six years, and having a higher level of trust about claims regarding the environmental impacts of products, are factors that are associated with a higher likelihood of investing in most of the water-efficient devices.

2. Research on the drivers of water conservation

Given the need for decision makers to develop policies that promote water conservation and protect the environment, it is important to identify the individual and household characteristics which affect water-saving behaviours and investment in water-efficient devices. There is a large literature focusing on the effects of water pricing, and the positive effect that this has on water conservation (for a review, see OECD, 2011). The results from the 2008 round of the OECD survey confirm these findings. Grafton et al. (2011) use data from ten OECD countries and find that volumetric water charges and higher water prices increase the probability of adopting water-saving behaviours. This is consistent with much of the previous literature.

More recently, attention has been given to the potential impact of environmental education and awareness on household behaviour. Nonetheless, the empirical analysis of behavioural attitudes and actions in water research has not been a common topic.

The links between residential characteristics, attitudinal characteristics, and environmental behaviours are complex but can be divided into three broad groups of variables: situational characteristics, environmental values, and psychological factors.

Situational characteristics are those that define a given personal situation, including individual demographic factors such as income, education, household size, residence type and ownership; environmental knowledge, e.g. general knowledge about the state of the environment and an awareness of environmental problems; and behavioural knowledge, that is, knowing how to implement one or other environmental behaviours. The second set of factors that have been found to affect environmental behaviour are *environmental values*, which refer to those underlying orientations held by individuals towards the physical environment, e.g. environmental concern and environmental attitudes. The third broad group of factors covers *psychological factors*, including individuals' personality characteristics and their perceptions towards the actions they are undertaking. De Young (1996) argued that people may adopt conservation behaviours because they enjoy doing so, and thus have an intrinsic motivation to undertake such behaviours, which is distinct from the strictly "environmental" motivation. Thus, personality traits, such as the extent to which people gain satisfaction and feel good from undertaking conservation behaviours, are important predictors of environmental behaviours.

In the academic research on water conservation, previous results about the influence of demographic factors on conservation behaviour were somewhat inconclusive. Hines et al. (1987) find that water conservation activities are associated with higher income groups; Berk et al. (1993) report that income and education have a positive effect on water conservation behaviour, whereas De Oliver (1999) reports that water conservation activities are associated with lower income and education.

Among the few studies that focus primarily on the effect of non-demographic factors on water-saving behaviours, Barr and Gilg (2006) use data from England and find that education level as well as being a member of community groups or environmental organisations have a positive effect on water-saving behaviours. Grafton et al. (2011) find that being a member or supporter of an environmental organisation and having a greener attitude towards the environment also increase the probability of undertaking most water-saving behaviours. The social norm of the respondents, represented by whether they had voted in local or national elections in the past six years, was also found to have a significant and positive effect on water-saving behaviours.

One of the non-price demand-side policies that can be used to promote water conservation is to encourage the installation of water-efficient devices in residential housing. Several countries have implemented rebate programmes for the installation of water-efficient equipment such as dual-flush toilets and water-efficient shower heads. However, owing to a lack of appropriate data on the adoption of water-efficient devices, very few studies have investigated the factors that encourage their adoption at a household level, and those that have, mainly controlled for socio-economic variables. The effect of attitudinal variables on households' adoption of water-efficient devices has been overlooked, except in a recent study by Millock and Nauges (2010) who found that environmental attitudes and ownership status are strong predictors of the adoption of water-efficient devices.

In order to better understand the factors affecting water-saving behaviours and the adoption of water-efficient devices by households, as well as factors driving satisfaction with the quality of tap water and the correlation between this satisfaction with primary sources of drinking water, this chapter simultaneously examines the effect of social-economic variables and attitudinal variables on water conservation behaviour in OECD member countries. The analysis offers policy recommendations to water authorities who wish to find effective tools to promote water conservation.

3. Overview

The data for this analysis came from the 2011 OECD Survey on Environmental Policy and Individual Behaviour Change (EPIC) which collected data at a household level across five environment-related areas: waste, transport, energy use, food, and water consumption. The survey was implemented using an internet-based questionnaire with some 12 000 respondents across eleven OECD countries: Australia, Canada, Chile, France, Israel, Korea, Japan, the Netherlands, Spain, Sweden and Switzerland. The survey collected information on each household's economic and demographic characteristics (income, age, gender, household size and composition, education, employment status, residence size, type of residence, and so on); attitudinal characteristics (environmental concerns, membership of a non-governmental organisation, participation in civil society, etc.); and policy-specific characteristics (such as water pricing structure, labelling schemes).

In the area of water consumption, respondents were asked a series of questions regarding their water-saving behaviours: turning off water while brushing teeth; plugging the sink when washing dishes; watering the garden in the coolest part of the day to save water; collecting rainwater/recycling waste water; and taking a shower instead of a bath. In addition, they were asked about the adoption of water-saving devices: low-volume or dual-flush toilets; water flow-restrictor taps/low-flow shower heads; and, using a water tank to collect rainwater. They were also asked about their level of satisfaction with the quality of tap water; their primary sources of drinking water (straight from the tap, purified/boiled tap water, etc.). Full descriptions and definitions of all the variables used in the analysis are provided in Appendix 5.A1.

Table 5.1 shows that among all respondents 73% of households face volumetric water charges with individual water metering. The lowest reported rates of volumetric charges are in water-abundant Canada and Sweden. Chile and Japan have the highest rates. However, almost 5% of respondents report that they do not know whether and how they are charged for water consumption – with this figure being highest in Canada (7.6%) and lowest in Chile (1.3%).

Table 5.1. **Water charge types (as percentage of total responses)**

	Individual volumetric water charge				Observations
	Yes	*No*	*Other charge*	*Don't know*	
Australia	74.8	15.2	3.0	7.0	996
Canada	36.1	52.0	4.4	7.6	1 122
Chile	90.6	4.8	3.3	1.3	1 027
France	81.2	11.0	4.9	2.9	1 227
Israel	84.4	6.2	4.3	5.1	1 168
Japan	92.3	4.2	2.0	1.5	1 043
Korea	81.9	3.7	11.2	3.2	1 116
Netherlands	86.7	5.1	2.5	5.7	1 301
Spain	82.1	7.0	7.6	3.3	1 101
Sweden	32.7	56.9	5.4	5.0	1 012
Switzerland	51.1	30.3	12.1	6.5	1 089
OECD(11)	***72.6***	***17.4***	***5.5***	***4.9***	***12 202***

A comparison between percentages of households facing volumetric water charges in 2011 and 2008 is shown in Figure 5.1. For all countries where the comparison was feasible, the percentage of households charged volumetrically for their water consumption was higher in 2011 than in 2008. However, it must be noted that the samples are different in the two rounds of the survey.

Figure 5.1. **Percentage of households facing a volumetric water charge, 2011 and 2008**

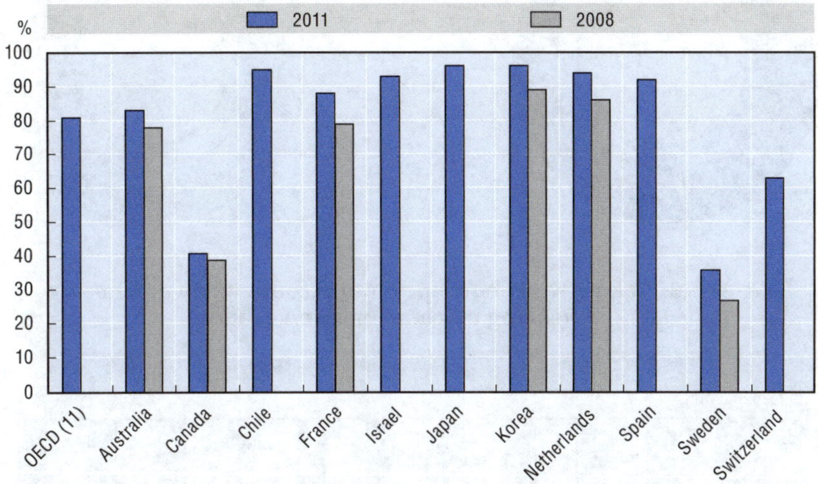

Information on the adoption of water-saving devices is given in Figures 5.2a and b which show that, among all respondents in the past ten years: 34% of households had invested in low-volume or dual-flush toilets. However, an additional 17% stated that their residence was already equipped with such a toilet and 13% that it was not feasible. The overwhelming majority of the latter are tenants. Of those who had invested in low-volume or dual-flush toilets, 13% had received financial support to that effect.

Figure 5.2 also provides data on investment in water-flow restrictor taps and water tanks. 41% report that they had invested in water flow restrictor taps or low flow shower heads and 12% have invested in water tanks to collect rainwater. Over 20% of those who had invested in water-flow restrictor taps responded that they had received financial support – i.e. over 1 000 households in the total sample. (see Figure 5.2B). The characteristics of the recipients are described in greater detail below.

Forty-seven per cent of respondents state that they take water efficiency into account when purchasing washing machines or dishwashers (Figure 5.3). France, Australia, the Netherlands and Spain have the highest percentages.

Figure 5.2. **Investments in appliances and receipt of financial support to make these investments (in percentage)**

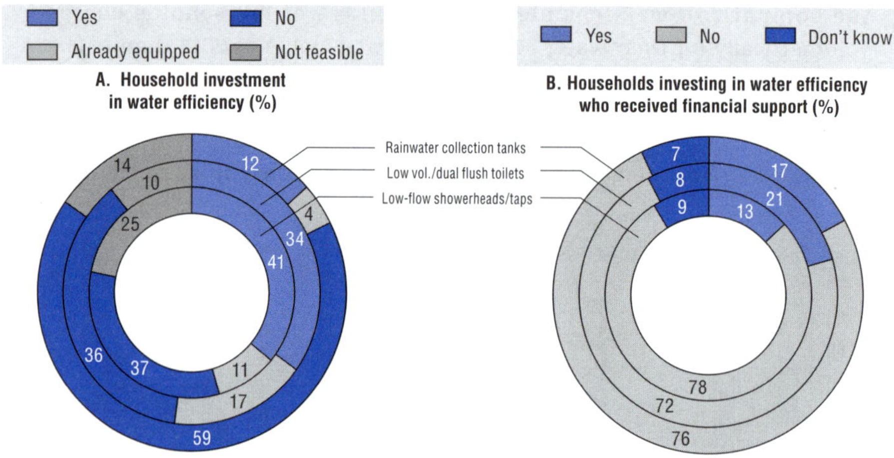

Figure 5.3. **Taking water efficiency into account when purchasing a washing machine or dishwasher**

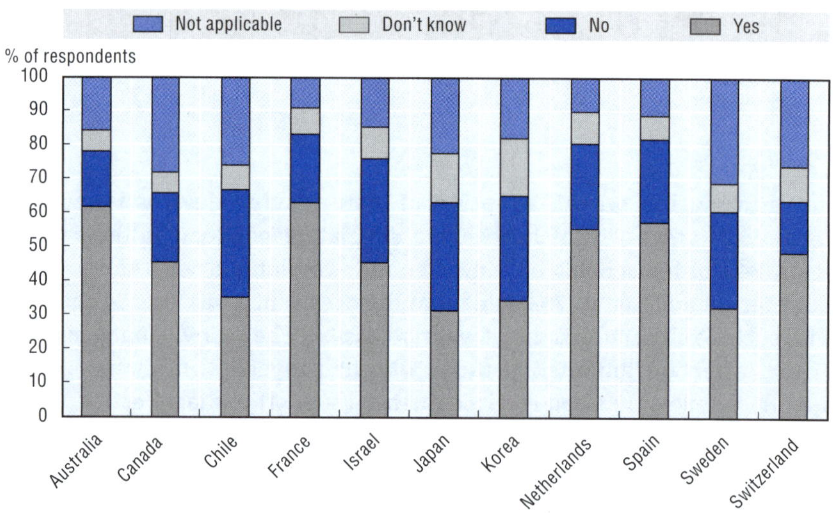

The figure for water-scarce Israel is surprisingly low. However, it must be borne in mind that almost 30% of the total sample state that they "don't know" whether they do so (perhaps. somebody else in the household is responsible for such purchases) or that it is not applicable (when they do not own a washing machine or dishwasher).

As noted, the survey considered five water-saving behaviours. A summary of the percentage of people classifying themselves through each of the possible responses (never, occasionally, often, always, and not applicable) is provided in Figure 5.4. It is important to note that structural and cultural factors can play an important role in affecting these behaviours. For instance, to the question about garden watering and rainwater collection, many respondents indicated that it is "not applicable". Moreover, it is not always clear that particular behaviours are water saving in all contexts (e.g. "take shower instead of bath"). Ongoing empirical work will help to tease out the context-specific factors which are at play.

Figure 5.4. **Frequency of undertaking water-saving behaviours**

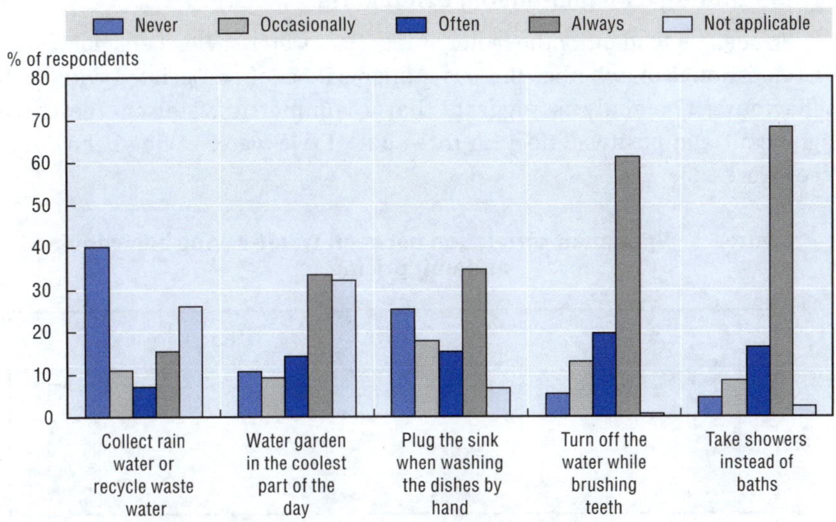

4. Analysis and results

This chapter focuses on descriptive statistics and bivariate relationships between variables of interest. The review of the data seeks to identify key relationships regarding the following policy-relevant questions:

1. How do water-saving behaviours and households' adoption of water-efficient equipment vary with demographic and socio-economic variables?
2. How do respondents' social norms, concerns for the environment, andgeneral attitudes towards the environment correlate with water conservation behaviours and investment in water-efficient equipment?
3. What is the effect of volumetric water charges on household water conservation behaviours and investment in water-efficient equipment?

4. Does the presence of water labelling schemes affect water-saving behaviours? What are the effects of label recognition, comprehension and trust?
5. Who invests in water-efficient appliances and devices? Is this affected by the presence of water labelling schemes? Who benefits from financial support measures to adopt water-efficient equipment?
6. Which households are more likely to be unsatisfied with the quality of their tap water? How does this correlate with households' primary sources of drinking water?

Subsequent work will seek to examine the underlying relationships in a more formal manner.

Factors that affect water-saving behaviours

In seeking to identify the factors that affect water-saving behaviours, the first question is on whether there is a financial benefit associated with such behaviours. The analysis suggests that a volumetric water charge has a significant and positive effect on three out of five water-saving behaviours (see Figure 5.5)

Figure 5.5. **Spearman correlation between water-saving behaviours and unit pricing**

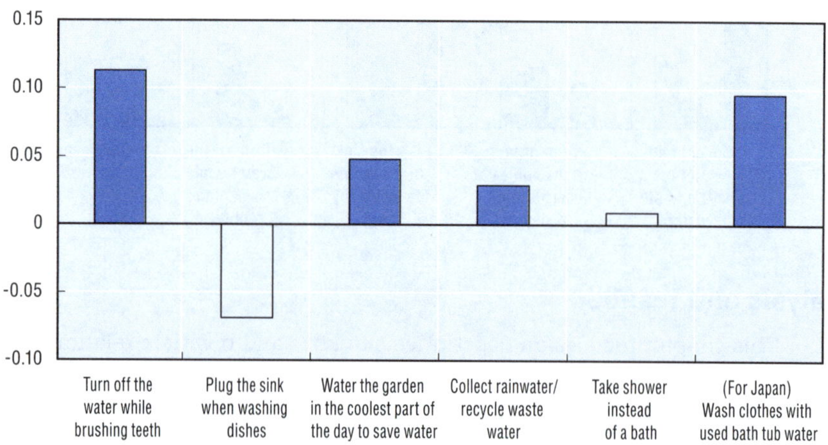

Note: The Spearman correlation is similar to a standard correlation (both measure how closely two variables move with each other), but the Spearman correlation is often used to examine the relationships between attitudinal data. Unfilled bars indicate no statistical significance at the 10% level. Sample does not include those who say the measure is not applicable.

The analysis shows an insignificant relationship for the effect of a volumetric water charge in two cases ("take shower instead of bath" and "plug the sink when washing dishes"). In the first case, this is hardly surprising since it is likely that preferences for one or the other means of washing as well as available facilities are the determinant factors. For the Japanese sample, a specific question was asked on whether clothes were washed with used bath or tub water, and the effect of volumetric charges is significant in this case.

Looking more closely at those behaviours for which there is an apparent relationship with unit water charging, one finds that in nine of eleven countries households facing unit water charges are slightly more likely to turn off water while brushing teeth, though the difference is quite small. Regarding outdoor water use and rainwater collection, Figures 5.6 and 5.7 show that in ten countries, households subject to unit water charges have adopted habits for more efficiently using outdoor water use – watering gardens in cooler parts of the day, collecting rainwater, and/or recycling wastewater. Chile, Australia and France are the countries that have the highest frequencies of "often" or "always" watering the garden in the coolest part of the day to save water. However, the effect of unit charging appears to be greatest in Chile, France and Spain. In ten of the eleven countries, those facing water charges are more likely to collect rainwater or recycle waste water, the exception being Chile. Australia and France are also among the countries that have the highest frequencies of collecting rainwater and recycling waste water. Switzerland, France and Korea are the countries where unit charging appears to have the greatest impact on behaviour.

Figure 5.6. **Relationship between watering the garden in the coolest part of the day to save water and unit water charge**

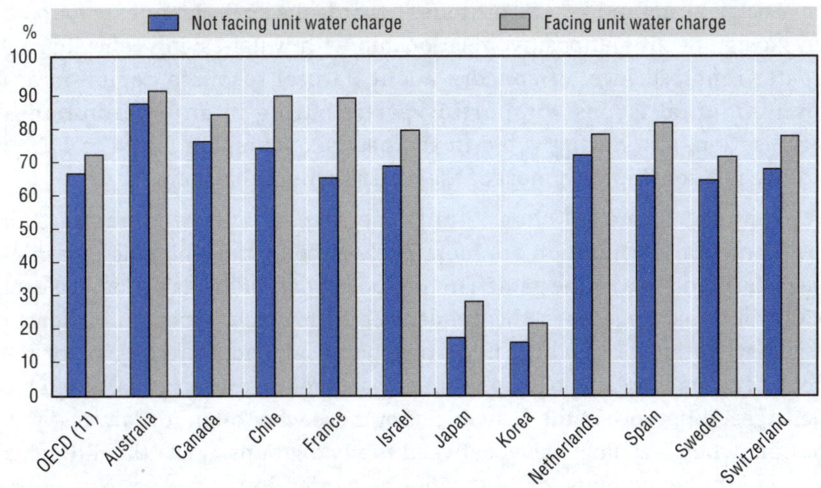

Note: "Not applicable" respondents are excluded from the sample.

Figure 5.7. **Relationship between collecting rainwater/recycling waste water and unit water charge**

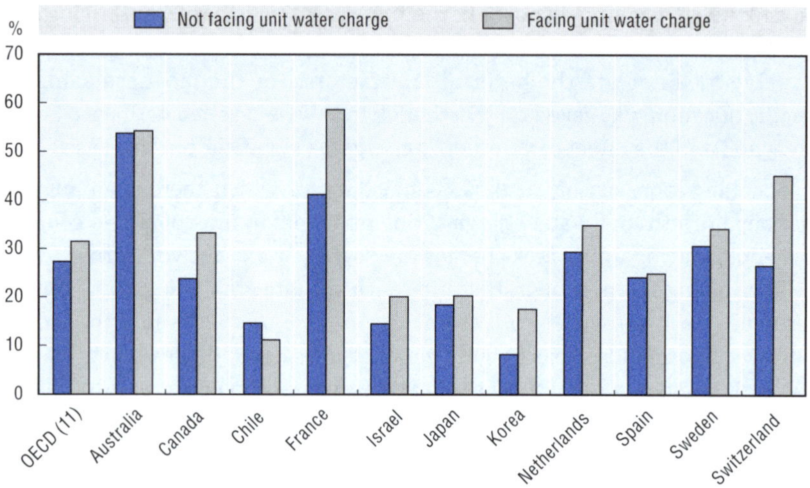

Notes: "Not applicable" respondents are excluded from the sample.

Table 5.2 reports the Spearman correlation coefficients between water-saving behaviours and economic, demographic and attitudinal characteristics. Among the demographic and socio-economic factors, household income and years of post-secondary education have a significant and negative relationship with three of the water-saving behaviours. The effects of the other variables are more mixed. In general, home ownership has a positive effect, and living in an urban location has a negative effect.

Social norms and general attitudes towards the environment are found to have a significant and positive relationship with water-saving behaviours. A greater reported level of concern about natural resource depletion and environmental issues, supporting/participating in an environmental organisation, and having voted in the past six years, are found to have a positive relationship with most of the water-saving behaviours.

Figures 5.8 and 5.9 show that those respondents who support an environmental organisation are more likely to "plug the sink when washing the dishes" and "water the garden in the coolest part of the day". For all eleven OECD countries considered, the frequencies with which respondents indicate that they "often" or "always" undertake such behaviour are higher for the group of respondents who support an environmental organisation. Spain has the largest difference in the frequencies of those who "often" or "always" plug the sink when washing dishes between the two groups, while the difference between the two groups for "watering the garden in the coolest part of the day" is largest in Israel and Sweden.

Table 5.2. **Correlation coefficients between water-saving behaviours and selected variables**

	Turn off the water while brushing teeth	Plug the sink when washing dishes	Water the garden in the coolest part of the day to save water	Collect rainwater/ recycle waste water	Take a shower instead of a bath	Wash clothes with used bath or tub water (for Japan)
Demographic and socio-economic variables						
Income	-0.014	-0.030**	-0.003	-0.042***	-0.028**	0.072
Gender (male = 1)	-0.075***	0.022*	-0.085***	-0.030**	-0.001	0.0003
Age	-0.076***	0.167***	0.104***	0.111***	0.0001	0.035
Household size	0.062***	-0.077***	0.007	-0.021	-0.015	0.241***
Years_post_school	0.006	-0.055***	0.007	-0.042***	-0.038***	-0.019
Housing attributes						
Owner_occupier	0.033***	0.057***	0.061***	0.080***	0.015	0.120***
Years_in_residence	-0.006	0.062***	0.066***	0.051***	-0.045***	-0.032
Urban_location	0.013	-0.123***	-0.109***	-0.161***	-0.022*	-0.044
Concerns, attitudes and norms						
Rank_env_concern	0.060***	0.001	-0.023	0.021	-0.035***	0.045
Cncrn_resource_depl	0.116***	-0.013	0.134***	0.031***	0.083***	0.046
Env_cncrn_index	0.131***	-0.010	0.124***	0.039***	0.095***	0.137***
Voter_dummy	0.028**	0.023*	-0.010	0.049***	0.002	0.053
Env_group supporter	0.053***	0.085***	0.133***	0.089***	0.030**	-0.014
Env_attid_index	0.111***	0.009	0.139***	-0.003	0.105***	0.052

Note: Spearman correlation. *, **, *** indicate statistical significance at the 10%, 5%, and 1% significance level, respectively. Sample does not include those who say the measure is not applicable.

Figure 5.8. **Relationship between plugging the sink when washing dishes and supporting/participating in an environmental organisation**

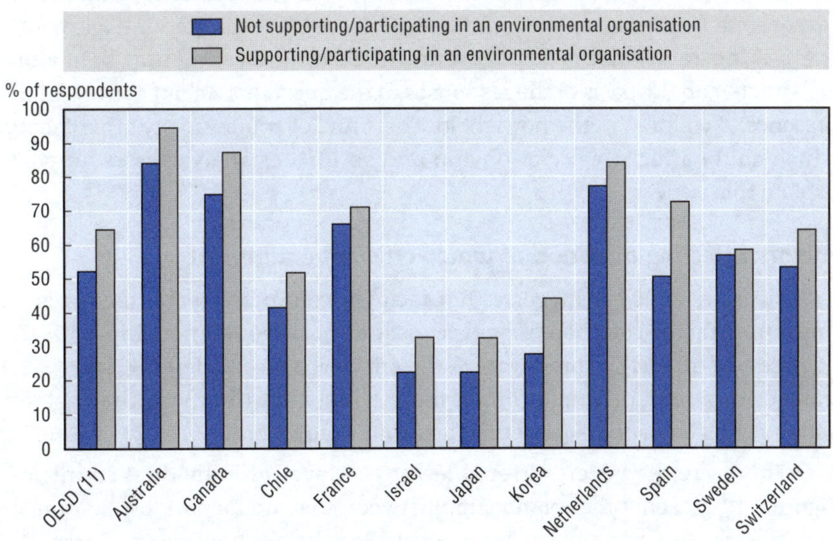

Note: "Not applicable" respondents are excluded from the sample.

Figure 5.9. **Relationship between watering the garden in the coolest part of the day to save water and supporting/participating in an environmental organisation**

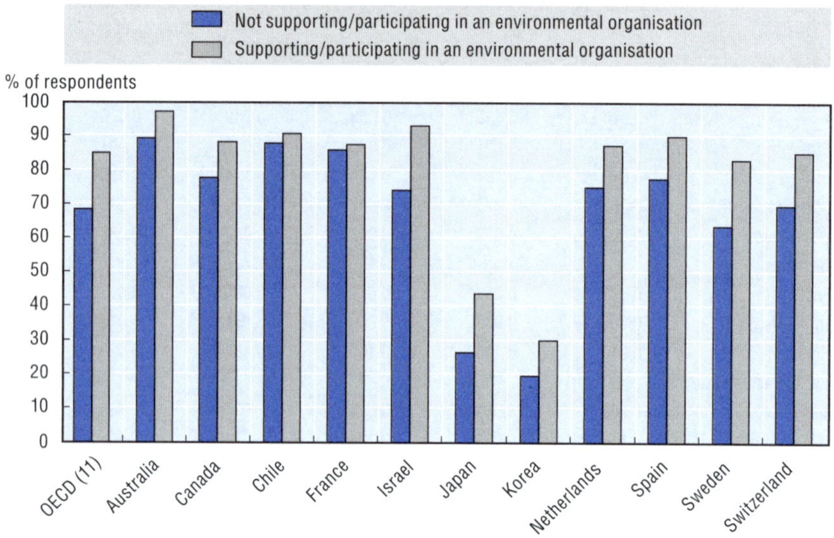

Note: "Not applicable" respondents are excluded from the sample

It is often argued that the presence of children can have a positive impact on more environmentally responsible behaviour. This could arise because of the role played by school curricula, encouraging greater environmental awareness. It could also be a consequence of a wish on the part of parents to demonstrate responsible behaviour to their children. However, as indicated in Figure 5.10, there is no relationship between an index of environmental behaviour constructed on the basis of the responses to the questions set out above and the presence of children in the household. There are, of course, many other factors which could affect this relationship and so further analysis is required to explore this issue.

Factors affecting adoption of water-efficient equipment

Bivariate analysis suggests that facing a volumetric water charge is an important factor that has a significant and positive relationship with the adoption of all of the three water-efficient devices considered. (Figure 5.11) Indeed, this relationship is much more consistent than was the case for water-saving behaviours presented above in Figure 5.5.

There are, however, differences at the level of individual countries. Figure 5.12 presents the relationship between the adoption of low-volume or dual-flush toilets and unit water charges. For all countries except Korea, the percentage of adoption of low-volume or dual-flush toilets is higher for

Figure 5.10. **Relationship between presence of children and index of water conservation behaviours**

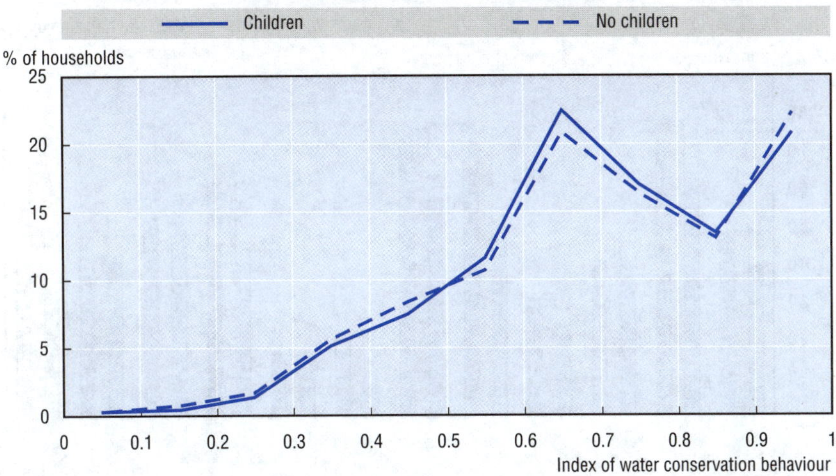

Note: The index of water conservation behaviour was created on the basis of the responses to "plug the sink", "turn off water while brushing teeth" and "take showers instead of baths". A value of 0 indicates never engaging in any of these behaviours, while a value of 1 corresponds to always adopting these behaviours.

Figure 5.11. **Correlation between water-saving investments and unit pricing**

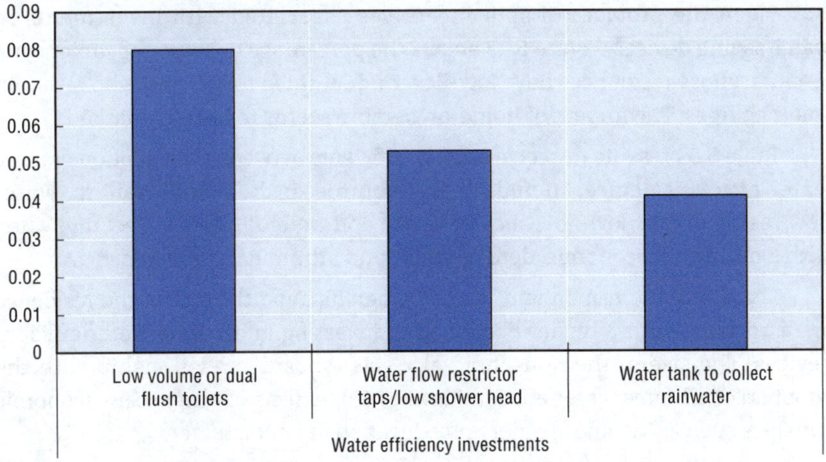

Note: Spearman correlation.

households facing unit water charges than for households not facing such charges. For Korea, the percentage is nearly the same between the two groups. The percentage of households investing in low-volume or dual-flush toilets is highest in Australia and Israel (about 73%), and lowest in Korea (17%). In a similar

Figure 5.12. **Relationship between adoption of low-volume or dual-flush toilets and unit water charges**

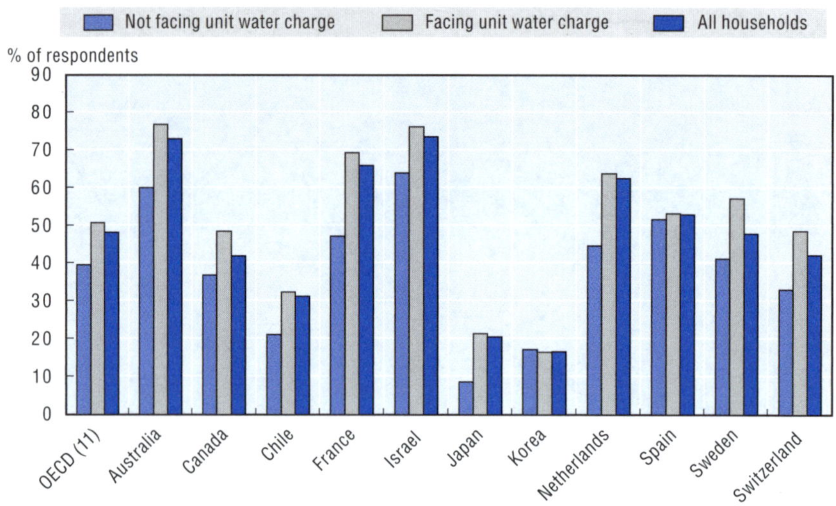

pattern, a positive relationship between the adoption of water flow-restrictor taps/low-flow shower heads and the adoption of water tanks to collect rainwater with unit water charges is present in all eleven OECD countries.

Given the problem of split incentives associated with investments in water-saving devices, Figure 5.13 shows the percentage of owner-occupiers and tenants whose residences have low-flow or dual-flush toilets and who face unit water charges. The impact of home-ownership seems to be very marked.

Table 5.3 presents the correlation coefficients between the adoption of three water-efficient device, including low-volume or dual-flush toilets, water flow-restrictor taps/low-flow shower heads, and water tanks to collect rainwater, with a number of economic, demographic, and attitudinal characteristics.

The age of the respondent, home ownership, and the size of the residence are associated with a higher likelihood of investing in all three water-efficient devices. The size of the household also has a positive relationship with the adoption of all three water-efficient devices. The effects of other socio-economic variables such as income, gender and education are not apparent.

Social norms and attitudinal characteristics have a significant and positive link with the adoption of all three water-efficient devices. A higher level of concern about natural resource depletion and environmental issues, supporting/participating in an environmental organisation, having voted in the past six years, and having a higher level of trust about claims regarding the environmental impacts of products, are factors that are associated with a higher likelihood of investing in most of the water-efficient devices.

Figure 5.13. **Investment in low-flow toilets and ownership status for those facing unit water charges**

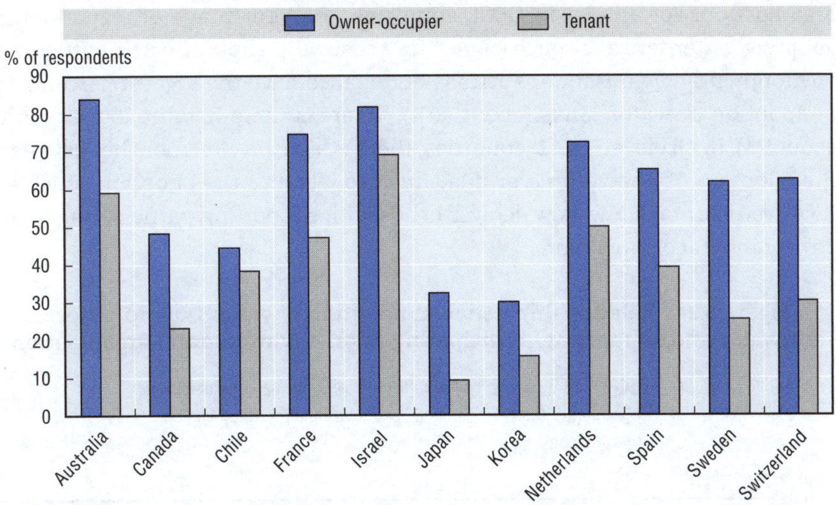

Note: Proportion of "yes" + "already equipped", over "yes" + "already equipped" + "no".

Table 5.3. **Correlation between investing in water-saving devices and selected variables**

	Low-volume or dual-flush toilets	Water flow-restrictor taps/low-flow shower head	Water tank to collect rainwater
Demographic and socio-economic variables			
Income	0.029**	0.006	-0.017
Gender (male = 1)	0.003	0.018	-0.019
Age	0.100***	0.070***	0.055***
Household size	0.019	0.053***	0.016
Years_post_school	0.018	0.012	-0.052***
Housing attributes			
Owner_occupier	0.132***	0.074***	0.084***
House_dummy	0.089***	0.075***	0.191***
Residence_size	0.139***	0.115***	0.125***
Concerns, attitudes and norms			
Cncrn_resource_depl	0.025**	0.044***	0.015
Env_cncrn_index	-0.007	0.041***	0.004
Trust_index	0.046***	0.082***	0.021
Voter_dummy	0.074***	0.073***	0.046***
Env_group supporter	0.093***	0.093***	0.077***
Env_attid_index	0.059***	0.072***	-0.008
Watr_bhv_index	0.188***	0.199***	0.350***

Note: Spearman correlation.

Figure 5.14 shows the relationship between the adoption of water tanks to collect rainwater and supporting/participating in an environmental organisation. For all eleven OECD countries, the percentage of respondents adopting water tanks is much higher for those who support/participate in an environmental organisation. Australia and France have the highest percentage of adoption of water tanks (about 40%), whereas Chile has the lowest (only about 5%). In all eleven OECD countries, there is a positive relationship between the adoption of low-volume or dual-flush toilets and the adoption of water flow-restrictor taps/low-flow shower heads with supporting/participating in an environmental organisation.

Figure 5.14. **Relationship between adoption of water tank to collect rainwater and supporting/participating in an environmental organisation**

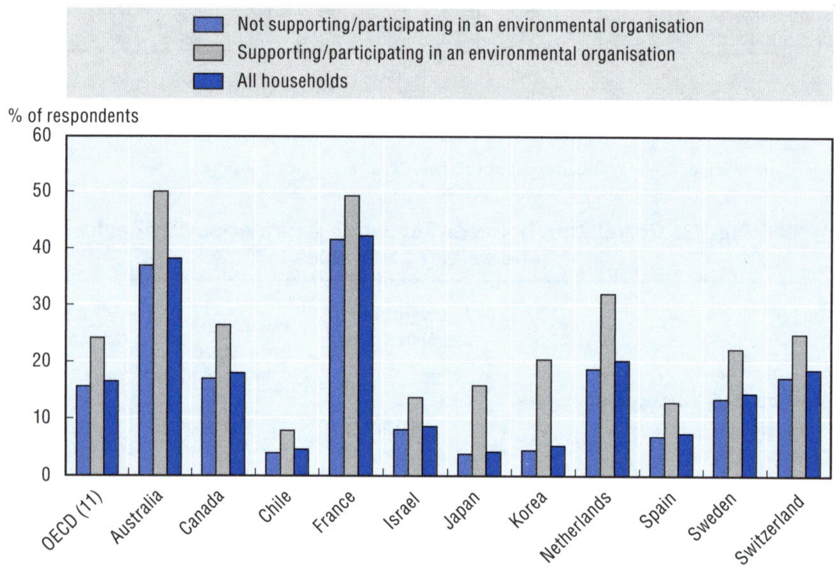

Table 5.4 presents the correlation between the adoption of the three water-efficient devices with labelling variables, for Australia, Israel and the Netherlands.[2] In Australia and the Netherlands, respondents' recognition of water labels has a significant and positive relationship with the adoption of all three water-efficient devices, although the effect of their recognition of water labels is positive but insignificant in Israel. However, the effect of understanding and trusting water labels on the adoption of water-efficient devices is less apparent from the analysis.

Respondents were also asked to indicate if they took water efficiency into account when purchasing washing machines or dishwashers. The data

Table 5.4. **Correlation between investing in water-saving devices and labelling**

		Recognise label	Understand label	Trust label
Low-volume or dual-flush toilets	OECD (3)	0.114***	0.021	-0.035
	Australia	0.094**	0.042	-0.019
	Israel	0.037	0.017	-0.051
	Netherlands	0.111***	0.001	-0.030
Water flow-restrictor taps/low-flow shower head	OECD (3)	0.073***	0.067**	0.038
	Australia	0.084**	0.050	0.017
	Israel	0.031	0.080***	0.006
	Netherlands	0.115***	0.076	0.136**
Water tank to collect rainwater	OECD (3)	0.096***	0.088***	0.071**
	Australia	0.107***	0.022	-0.025
	Israel	0.017	0.013	0.050
	Netherlands	0.051*	0.132**	0.127*

Note: Spearman correlation. *, **, *** indicate statistical significance at the 10%, 5%, and 1% significance levels, respectively.

presented in Figure 5.15 indicate that recognition has a positive effect. Not surprisingly, "use" of the label also correlates with the responses provided to this question.

Figure 5.15. **Relationship between taking water efficiency into account when purchasing appliances and respondents' recognition and use of water labels**

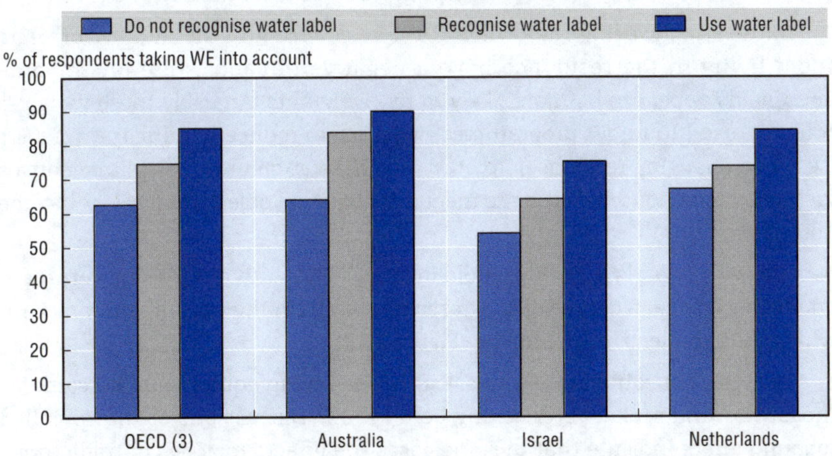

Figure 5.16 shows the correlation between receiving government financial support for different water-efficient devices and a set of explanatory variables. The variable takes a value of one if grants have been received for any of the three types of equipment in which respondents may have invested. The

Figure 5.16. **Correlation between receiving water conservation grants and selected variables**

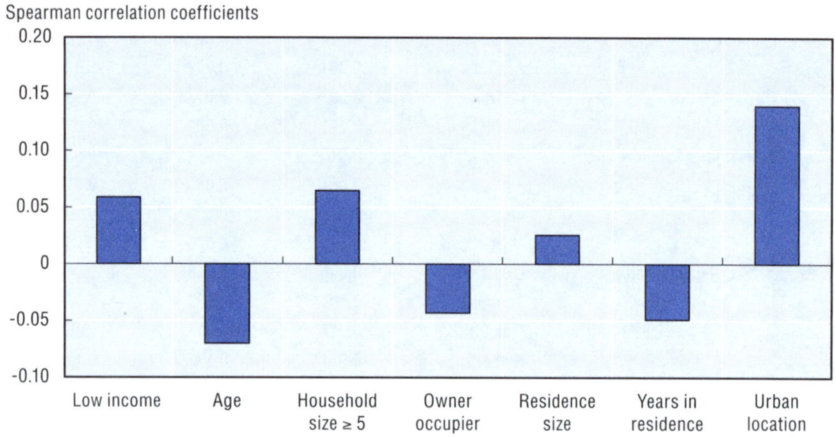

Note: Spearman correlation.

analysis suggests that low income, large household size, large residence size, and living in urban areas have a significant and positive relationship with the likelihood of receiving water conservation grants. The age of the respondent has a significant and negative relationship with the receipt of water conservation grants. The analysis finds no significant effect of gender and employment status of the respondent on the receipt of water conservation grants.

Interestingly, being the owner of the residence and the number of years spent living in the residence have a negative impact, while low-income households appear to be more likely to receive grants. Arguably these variables could be used to target programmes, whether to reduce administrative costs (i.e. when changing residence), to overcome market failures (i.e. split incentives for low-income households), or to meet distributional objectives (i.e. low-income households).

Closer comparison of household income levels with receipt of government grants does reveal a degree of progressivity, except in the case of rain tanks, for which the sample is much smaller (Figure 5.17).

Are the households that invest in water-saving equipment the same as those who undertake water-saving behaviours? Discussions of the so-called rebound effect indicate that in some cases the effects may be contradictory – at least in part. For instance, those who invest in water-restrictor taps may be less likely to conserve water by turning off the water while brushing their teeth. However, Figure 5.18 shows a positive relationship between water conservation investments and behaviour.

Figure 5.17. **Relationship between income and receipt of grants for different water-efficient devices**

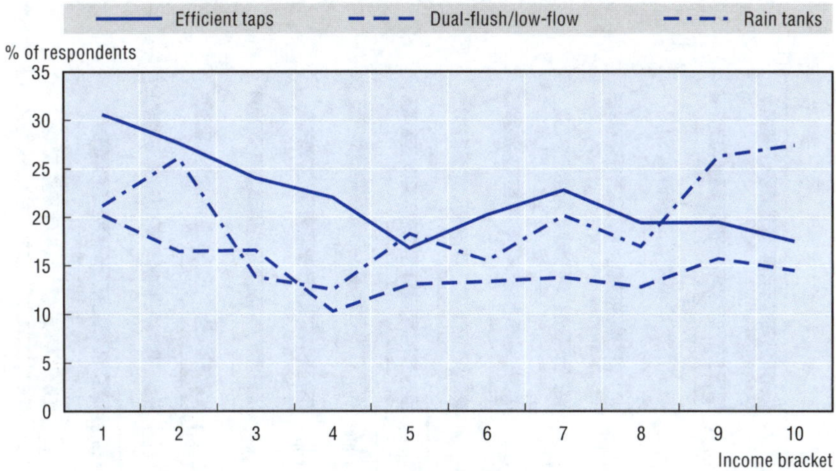

Note: Brackets differ by country, representing (approximately) deciles. "Don't know" and "Prefer not to answer" represent approximately 15% of the total sample.

Figure 5.18. **Relationship between investment in water-conservation devices and index of water-saving behaviour**

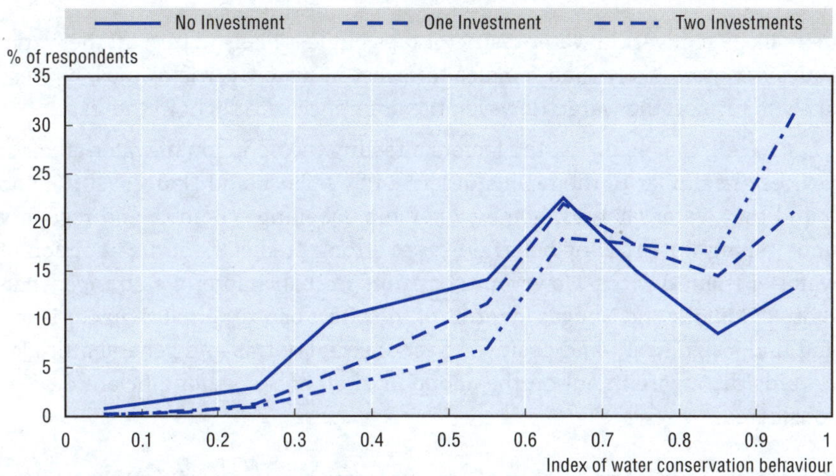

Note: The index of water conservation behaviour was created on the basis of the responses to "plug the sink", "turn off water while brushing teeth" and "take showers instead of baths". A value of 0 indicates never engaging in any of these behaviours, while a value of 1 corresponds to always adopting these behaviours.

Figure 5.19 presents a comparison of the demographic characteristics of those who often or always turn off the water while brushing their teeth and invest in water-restrictor taps. It is interesting to see that older respondents are

Figure 5.19. **Water-saving behaviour and investments, and demographic characteristics**

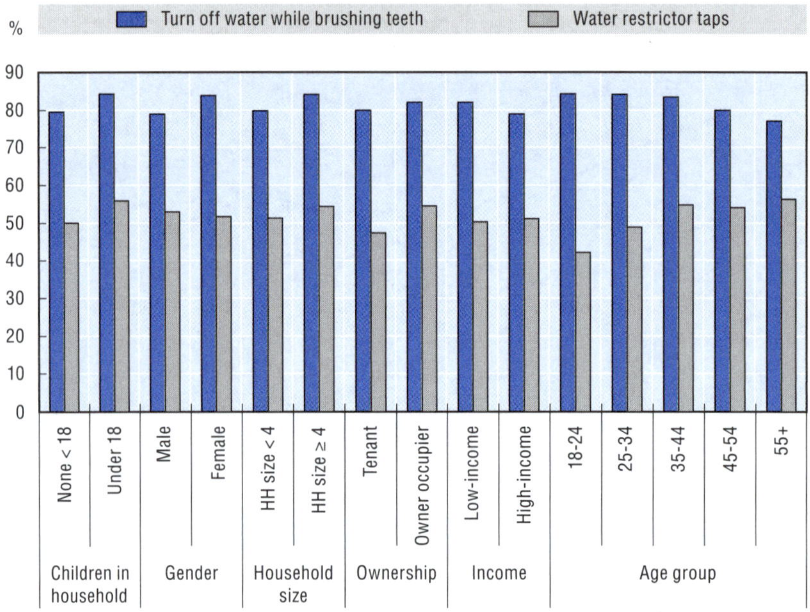

more likely to invest in water-restrictor taps but less likely to turn off the water. Males are more likely than females to invest in water-restrictor taps, but less likely to turn off the water. However, the differences are slight in this case.

Overall, the results on the factors affecting the adoption of water-efficient devices are similar to the results described by Millock and Nauges (2010) who found that water charges, home ownership, labelling schemes, and regularly undertaking water-saving behaviours, are strong predictors of the adoption of water-efficient devices. However, the results in this chapter are stronger than those of Millock and Nauges in terms of the effect of attitudinal characteristics; and find a significant and positive effect of social norms and general attitudes towards the environment on the adoption of all of the water-efficient devices considered.

Satisfaction with the quality of tap water and sources of drinking water

Respondents were asked to indicate their usual source of drinking water (unpurified tap, purified tap, bottled, natural source) as well as their level of satisfaction with the taste and health impacts of tap water (Figures 5.20A and B). Further investigation was carried out to determine which households are most likely to be dissatisfied with the quality of their tap water and the impact this has on their primary sources of drinking water.

Figure 5.20. **Relationship between satisfaction with tap water and source of drinking water**

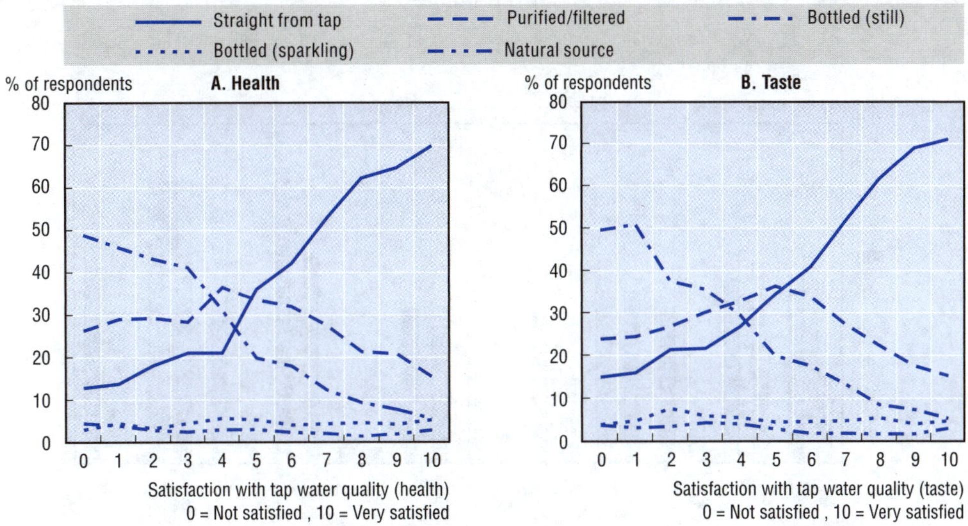

Those with high reported levels of satisfaction drink water straight from the tap. At low levels, there are high percentages of those who drink purified or bottled water. However, bottled water consumption decreases with satisfaction, while for purified water it increases and then decreases. This indicates that the two strategies are differently attractive depending upon the degree of dissatisfaction with unfiltered tap water. Both sparkling bottled water and reliance upon natural sources appear to be unrelated with the level of satisfaction with tap water quality.

Figure 5.21 further illustrates the positive relationship between drinking tap water and satisfaction with its taste. For all eleven OECD countries considered, the percentage of households drinking water straight from the tap is higher for those who are satisfied with its taste. However, the percentage of households drinking water straight from the tap varies widely. It is highest in the Netherlands and Sweden (about 87%) and lowest in Korea (about 5%).

Figure 5.22 presents the correlation between satisfaction with the quality of tap water and selected demographic, socio-economic and attitudinal variables. The results indicate that: people who are more satisfied with their life; people with a greener attitude toward the environment; and people who are more satisfied with the quality of their local natural water, are more likely to be satisfied with its taste and health impacts. Among the socio-economic characteristics, being a home-owner and older respondents are more likely to be satisfied with the quality of their tap water. On the other hand, households with a greater number of reported medical conditions within the household, and households with children under five years of age are more likely to be

Figure 5.21. **Relationship between drinking tap water and satisfaction with its taste**

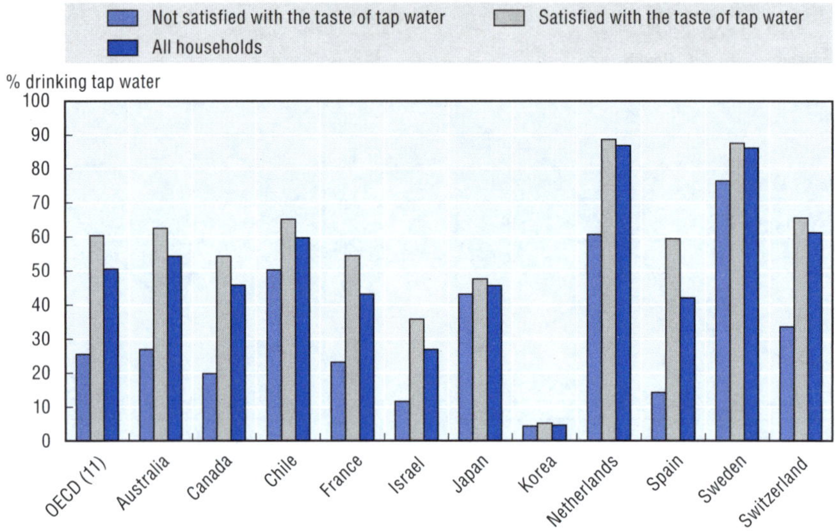

Note: In this case, those who respond 6 or over on a 10-point scale of satisfaction are considered to be "satisfied".

dissatisfied with the taste and health impacts of their tap water. (See Table 5.A1.2 in Appendix 5.A1 for full results. Table 5.A1.2 also provides data on the relationship between demographic characteristics and satisfaction with tap water.)

Figure 5.23 shows the correlation between drinking tap water and a set of explanatory variables. The data indicate that people who are more satisfied with the quality of their local natural water and older people are more likely to drink water straight from the tap instead of drinking water from other sources such as purified, boiled and bottled water. Households with children under five years old, being a home-owner, and those living in an urban area are less likely to drink water straight from the tap. Income and environmental attitudes do not have an impact. (See Table 5.A1.3 in Appendix 5.A1 for full results.)

An analysis between drinking bottled water and a set of explanatory variables shows that: people with a higher ranking of environmental concerns, people having voted in the past six years, and people who are satisfied with the quality of their local natural water, are less likely to drink bottled water. On the other hand, households with a greater number of medical conditions within the household and households with children under five years old are more likely to drink bottled water. By contrast to the result from the 2008 OECD survey (OECD, 2011), this analysis finds no significant effect of income and unit water charges on the likelihood of drinking bottled water.

Figure 5.22. **Relationship between satisfaction with tap water and demographic characteristics**

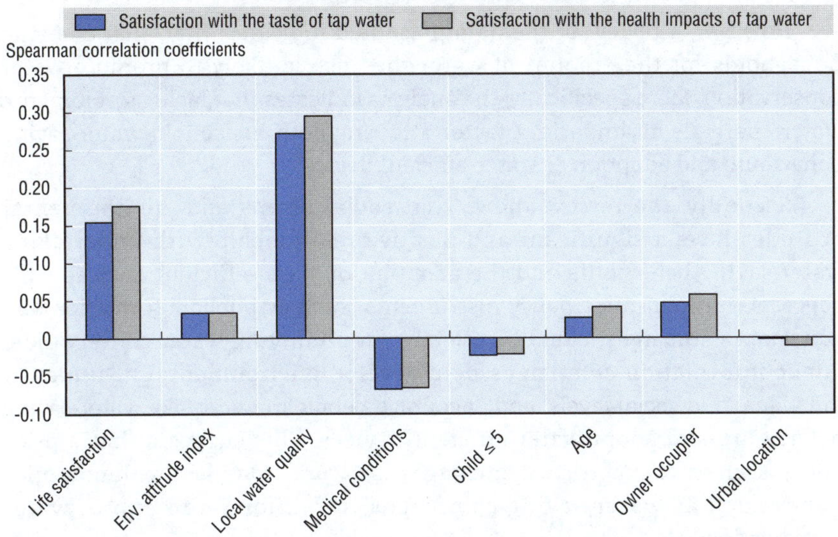

Note: Spearman correlation.

Figure 5.23. **Relationship between drinking tap water and demographic characteristics**

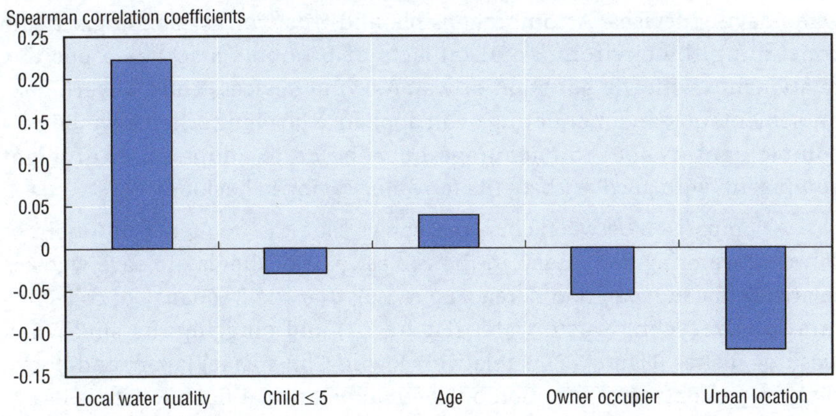

Note: Spearman correlation.

5. Preliminary policy implications

The results presented here are primarily descriptive in nature, showing the correlation between different socio-economic, demographic and attitudinal variables on water efficiency and the decision to use water from different sources. It is beyond the scope of this chapter to draw policy conclusions since the complex relationship between different sets of variables indicates the importance of undertaking more formal analysis. Follow-up data analysis is foreseen.

However, a review of the data presented indicates, first, that charging households for the amount of water they use is likely to promote water conservation. More specifically, the analysis indicates that facing a volumetric water charge is an important factor affecting both household water-saving behaviours and adoption of water-efficient devices.

Secondly, the review shows that social norms and environmental attitudes have a significant and positive relationship with undertaking water-saving behaviours and the adoption of water-efficient devices. This suggests that non-price policy instruments, such as public information and education campaigns, could be effective in promoting water conservation. Public information campaigns help develop environmental attitudes at individual and social levels, and encourage people to undertake water-saving behaviours and adopt water-efficient devices. This suggests that a price policy such as volumetric water charges can work in tandem with non-price policy such as water saving campaigns to reinforce the aim of water conservation.

Thirdly, labelling of water appliances is shown to be effective in Australia, Israel, and the Netherlands (Figure 5.15 and Table 5.4). Respondents' familiarity with water efficiency labels is correlated with the adoption of three water-saving devices. A comprehensible and trustworthy labelling scheme concerning the environmental impacts of products also has a positive relationship with the adoption of water-saving devices. Thus, government implementation and monitoring of an appliance labelling scheme would be a complementary tool to encourage households to adopt water-efficient equipment, with ancillary benefits for water-saving behaviour.

A comparison between countries shows that: the frequency of "often" or "always" watering the garden in the coolest part of the day to save water is relatively low in Japan and Korea (Figure 5.9); that the frequency of collecting rainwater/recycling waste water (Figure 5.7) and plugging the sink when washing dishes (Figure 5.8) is relatively low in Chile, Israel Japan, and Korea; that the percentage of adoption of low-volume or dual-flush toilets is low in Chile, Japan and Korea (Figure 5.12);[3] and that the percentage of adoption of water tanks to collect rainwater is low in Chile, Israel, Japan, Korea, and Spain (Figure 5.14).

Fourthly, one of the key uncertainties over the use of grants to encourage investment in water-efficient devices is that the recipients may be predominately households who would have invested in such devices without any government financial help. Thus, grants should be most effective if they are targeted on groups who are least likely to invest in water-saving devices.

The survey results show that those who are more likely to invest in most of the water-saving devices (aside from those with strong environmental attitudes and volumetric charges) are larger households, larger residences, and home-owners. There are good economic reasons for some of these findings – i.e. household-level economies of scale may explain why larger households are more likely to undertake such investments. Similarly, home-owners are better able to recover the benefits from their investments. There is some evidence that low-income households are favoured with such grants, but the degree of progressivity across the full range of income categories is less clear.

Fifthly, the survey finds that people with a greener attitude towards the environment, a higher level of satisfaction with the quality of their local natural water, and having voted in the past six years are more likely to be satisfied with the quality of their tap water. The survey also finds that households with a higher level of satisfaction with the quality of their tap water are more likely to drink water from the tap. This suggests that increasing the quality of tap water and public education campaigns about the importance of water and the environment can encourage people to drink tap water instead of drinking water from other sources such as purified, boiled and bottled water (and other non-alcoholic beverages in France).

To better understand the role of policy instruments in promoting water conservation, Figure 5.24 summarises the responses of people to the level of importance of each of the factors that encourage them to reduce their water consumption. These results show that, in the opinion of households, the availability of "less expensive water-efficient equipment", "easier identification of water-efficient appliances", the provision of "practical information on things you can do to save water at home" and "more information on water consumption by my household" are the most important factors that encourage people to reduce their water consumption. More than 80% of the respondents rate them as either fairly important or very important. Other factors, including "more information on environmental impact of water consumption", "finding my household uses more water than similar households" and "higher water price" are found to be relatively less important. Clearly, there may be an important strategic bias in the responses to the latter question, particularly in light of the overwhelming evidence in the literature on the positive impacts of water pricing on water conservation.

These results justify the importance of both price and non-price policies in encouraging people to reduce water consumption. In particular, public information and education campaigns that provide people with more information on the environmental impact of water consumption, the things they can do to save water at home, and the ability to identify water-efficient equipment; a clear water billing mechanism that provides households with accessible information on their water consumption and charges; and

Figure 5.24. **Importance level of factors that encourage people to reduce water use**

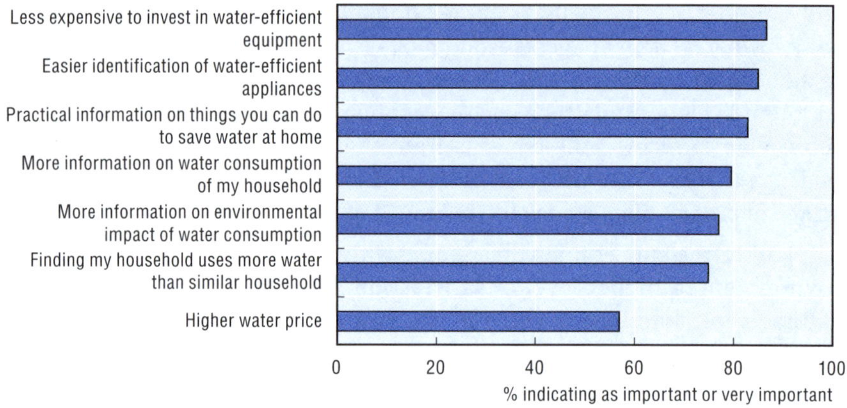

approaches for the adoption of water-efficient equipment that make investment in water-efficient equipment less expensive, would be effective in encouraging households to reduce water consumption.

6. Conclusions

Using a common survey instrument that collected data from around 12 000 people across eleven OECD countries, it is found that volumetric charges have an impact on water conservation behaviour and investments. The findings suggest that facing a volumetric water charge is an important factor affecting both household water-saving behaviours and adoption of water-efficient devices. In addition, labelling of water appliances correlates with water conservation behaviour. Recognition, comprehension and trust influence water-related decisions, indicating that government regulation and monitoring of labelling schemes that increase recognition, comprehensibility and trust will encourage households to adopt water-efficient devices and undertake water-saving behaviours.

The survey finds that the effectiveness of grants for investments in water-saving devices could be increased by targeting those households that are least likely to invest without financial incentives, such as low-income households and those living in rented properties. However, given the administrative cost and the role of other factors in the take-up of grants, more work is required to identify the efficiency of targeting programmes across different variables. Finally, the chapter finds that improving the quality of tap water, in combination with raising awareness of water issues through public information campaigns, can encourage households to move away from bottled or boiled water as their primary source of water and instead use more environment-friendly tap water.

In all areas, social and environmental norms and attitudes correlate with water conservation behaviour. In addition, different demographic and socio-economic groups appear to behave differently with respect to water use. However, it is important to disentangle the relationship between policies, environmental and social attitudes, demographic characteristics and behaviour. More formal multivariate empirical analysis is under way, with a view towards providing policy guidance.

Notes

1. This chapter focuses on households' direct use of water. Households also consume significant amounts of "virtual water" through consumption of good, such as food, paper and textiles, whose production uses large amounts of water (*http://www.virtualwater.eu/*).
2. These are the countries for which water efficiency labels were included in the survey questions.
3. Note that "standard" toilets in the different countries may have very different volumes of flow. As a consequence, responses to this question must be understood in light of very different baselines.

References

Arbués, F., M.A. García-Valina and R. Martínez-Espineira (2003), "Estimation of Residential Water Demand: a state-of-the-art review", *Journal of Socio-Economics*, Vol. 32, pp. 81-102.

Barr, S. and A. Gilg (2006), "Behavioural Attitudes towards Water Saving? Evidence from a Study of Environmental Actions", *Ecological Economics*, Vol. 57, pp. 400-414.

Berk, R.A. et al. (1993), "Measuring the impact of water conservation campaigns in California", *Climatic Change*, Vol. 24, pp. 233-248.

De Young, R. (1996). "Some psychological aspects of reduced consumption behaviour: the role of intrinsic motivation and competence motivation", *Environment and Behaviour*, Vol. 28, pp. 358-409.

De Oliver, M. (1999). "Attitudes and inaction: a case study of the manifest demographics of urban water conservation", *Environment and Behaviour*, Vol. 31, pp. 372-394.

Grafton, R.Q. et al. (2011), "Determinants of Residential Water Consumption: Evidence and Analysis from a 10-country Household Survey", *Water Resources Research*, Vol. 47, No. W08537, doi: 10.1029/2010WR009685.

Hines, J.M., H.R. Hungerford and A.N. Tomera (1987), "Analysis and synthesis of research on responsible.

Millock, K. and C. Nauges (2010), "Household Adoption of Water-Efficient Equipment: The Role of Social-Economic Factors, Environmental Attitudes and Policy", *Environmental Resource Economics*, Vol. 46, pp. 539-565.

OECD (2011), Greening Household Behaviour: The Role of Public Policy, OECD Publishing. doi: *10.1787/9789264096875-en*.

Terrebonne, R.P. (2005), Residential Water Demand Management Programs: *A Selected Review of the Literature*, Water Policy Working Paper 2005/02.

Van Den Bergh, J. (2008), "Environmental regulation of households: an empirical review of economic and psychological factors", *Ecological Economics*, Vol. 66, pp. 559-574.

APPENDIX 5.A1

Definition of variables

Unit_Charge:	Dummy = 1 if a household is charged according to how much water it uses, = 0 if a household is not charged or charged with flat fee (from Q91 in Annex A).
Income	Is the household income after tax (thousands of EUR/year) (from Q13).
Low_Income	Dummy = 1 for low-income group, i.e. for households in the two lowest income deciles in the survey; = 0 if otherwise (from Q13).
Income_Percptn	Dummy = 1 if the respondents described that they were living comfortably or very comfortably with their current income; = 0 if otherwise (from Q14).
Gender	Dummy = 1 if the respondent is a male; = 0 for a female (from Q3).
Age	Is the age of the respondent (years) (from Q4).
HH_Size	Is the number of people in the household (from Q5).
HHSize ≥ 5	Dummy = 1 if the household has 5 or more members; = 0 if otherwise (from Q5).
Child ≤ 5	Dummy = 1 if the household has a child under 5 years old; = 0 if not (from Q5).
Years_Post_Schol	Is the years of education the respondent completed after high school (from Q9).
Job	Dummy = 1 if the respondent is an employee, a student or self-employed, =0 if otherwise (from Q10).
Owner_Occupier	Dummy = 1 if the respondent is the owner of the residence; = 0 if not (from Q15).
House_Dummy	Dummy = 1 if the residence is a detached or semi-detached house; = 0 if otherwise (from Q16).
Residence_Size	Size of residence (m^2) (from Q17).
Years_in_Residence	Number of years the respondent has lived in the residence (from Q19).
Urban_Location	Dummy = 1 if the residence is being located in an urban or suburban area; = 0 if otherwise (from Q18).
Life_Satisfaction	Reflects level of satisfaction of the respondents with their life; values 1 to 10, a higher value means more satisfied (from Q21).
Rank_Env_Cncrn	Ranking of environmental concerns (from Q22). Values 1 to 6; a higher value means a higher ranking of environmental concerns.
Cncrn_Resource_Depl	Reflects concerns about natural resource depletion (forest, water, energy); a higher value means more concern (from Q23).
Env_Cncrn_Index	Reflects concerns about environmental issues; a higher value means more concern about the environment (from Q23).
Voter_Dummy	Dummy = 1 if the respondent has voted in local or national elections in the past six years; = 0 if otherwise (from Q24).
Env_Group_Supporter	Dummy = 1 if supporting or participating in activities of an environmental organisation; = 0 if otherwise (from Q25).
Env_Attid_Index	Reflects attitudes towards the environment; a higher value means a greener attitude towards the environment (from Q26).

EQ_Water	Reflects level of satisfaction of respondents with their local natural water, values 1 to 4; a higher value means more satisfied (from Q28).
Policy_Index_Vehicle	Reflects level of supporting government actions to reduce motor vehicle CO_2 emissions; a higher value means more support (from Q64).
Policy_Index_Waste	Reflects level of supporting government actions to reduce household waste generation; a higher value means more support (from Q42).
Satisfy_Health	Reflects level of satisfaction with the health impacts of tap water, values from 1 to 10; a higher value means more satisfied (from Q98).
Satisfy_Taste	Reflects level of satisfaction with the taste of tap water, values from 1 to 10; a higher value means more satisfied (from Q98).
Watr_Bhv_Index	Reflects a habit of undertaking water-saving behaviours; a higher value means undertaking water-saving behaviours more frequently (from Q92).
Trust_Index	Reflects level of trust on claims about environmental impacts of products; a higher value means a higher level of trust (from Q27).
Recognise_Label	Dummy = 1 if the respondent recognises water label; = 0 if otherwise.
Trust_Label	Dummy = 1 if the respondent trusts water label; = 0 if otherwise.
Use_Label	Dummy = 1 if the respondent uses water label; = 0 if otherwise.
Med_conds	Reflects the number of medical conditions in the household.
Satisfy_Taste	Dummy = 1 if the respondents are satisfied with the taste of their tap water, i.e. if *satisfy_taste* takes values from 6 to 10; = 0 if otherwise (from Q98).
Satisfy_Health_	Dummy = 1 if the respondents are satisfied with the health impacts of their tap water, i.e. if *satisfy_health* takes values from 6 to 10; = 0 if otherwise (from Q98).

Table 5.A1.1. **Correlation between receiving water conservation grants and selected variables**

	Receiving water conservation grants
lowincome	0.059***
highincome	-0.023
gender	0.008
Age	-0.070***
Hhsize_34	-0.007
Hhsize_5+	0.065***
Employed	0.012
home_owner	-0.043***
house_dummy	-0.084***
residence_size	0.026**
years_in_residence	-0.049***
urban_location	0.140***

Note: Spearman correlation.

Table 5.A1.2. Correlation between satisfaction with quality of tap water and selected variables

	Satisfied with the taste of tap water	Satisfied with the health impacts of tap water
Life_Satisfaction	0.155***	0.175***
Env_Cncrn_Index	-0.016	-0.005
Env_Attid_Index	0.033***	0.034***
EQ_Water	0.271***	0.294***
Med_Conds	-0.068***	-0.067***
Child_≤5	-0.024*	-0.022*
Gender (male=1)	0.025**	0.008
Age	0.027**	0.041***
HH_size	-0.007	-0.010
Owner_Occupier	0.046***	0.056***
Urban_Location	-0.002	-0.011

Note: Spearman correlation.

Table 5.A1.3. Correlation between drinking tap water and selected variables

	Drinking tap water
Unit_Charge	-0.080***
Income_Category	-0.014
Satisfy_Taste	0.317***
Satisfy_Health	0.310***
EQ_Water	0.223***
Child_≤5	-0.029**
Rank_EnvCncrn	0.007
Voter_Dummy	-0.015
Env_Attid_Index	-0.001
Gender (male=1)	0.013
Age	0.039***
Owner_Occupier	-0.056***
Urban_Location	-0.120***

Note: Spearman correlation.

Greening Household Behaviour
Overview from the 2011 Survey

Chapter 6

Household behaviour and food consumption

by
Katrin Millock *and* Céline Nauges*

This chapter looks at the impact of instruments directly targeting consumers' choices concerning food consumption, such as organic labelling and raising awareness through public information campaigns. It provides a better understanding of the main determinants for consuming organic food and products that take animal welfare into account, and examines how much more households are willing to pay for these products.

* CNRS-Centre d'Économie de la Sorbonne and Paris School of Economics, France, and Céline Nauges, The University of Queensland, Australia.

The statistical data for Israel are supplied by and under the responsibility of the relevant Israeli authorities. The use of such data by the OECD is without prejudice to the status of the Golan Heights, East Jerusalem and Israeli settlements in the West Bank under the terms of international law.

1. Introduction

In a context of growing population, rising incomes and lifestyle changes in large, densely populated countries, food production and consumption exert a growing pressure on the environment. At the same time, a number of food scares following outbreaks of BSE (mad-cow disease), dioxin-contaminated chicken, listeria and salmonella contamination have raised consumers' concern about food quality. Concerns about the environment and about product safety and health have induced some people to change their consumption patterns towards more environment-friendly products, including organic food.

Whereas some characteristics of organic food are recognised through experience (such as freshness and taste), environment-friendliness or health benefits are credence attributes in the sense that they cannot be ascertained directly by the customer who must instead rely only on certification or labelling. Consumers' socio-economic characteristics as well as attitudes are also known to influence organic food purchase.

A better understanding of the main determinants of consumers' behaviour with respect to certain food attributes is important when designing policies to raise consumer awareness through organic food labelling or the provision of public information and education campaigns. This chapter examines what drives organic food consumption and also some insights into household attitudes towards products taking animal welfare into account.

Drawing upon observations from over 12 000 households in eleven countries, the OECD Survey on Environmental Policy and Individual Behaviour Change (EPIC) provides insights into key issues related to the role of policy contexts, including:

- What encourages consumption of organic foods and are there significant differences across different household groups?
- To what extent do general households' attitudes towards the environment affect organic food consumption and consumption of food that takes animal welfare into account?
- What factors affect other aspects of food consumption such as the generation of food waste and "food miles"?
- How much are households willing to pay as a price premium to purchase organic foods? Does willingness-to-pay (WTP) vary significantly across household groups?

- How effective is labelling for organic food? For which type of household? Which labels are the more effective at inducing organic food consumption?

The following sections will analyse successively organic food consumption and its main determinants; the factors that encourage consumption; the associated impacts of food consumption such as food waste and food miles; the role of labelling; the willingness to pay a premium for organic food and for meat and poultry taking animal welfare into account. The last section is a summary of the main conclusions.

A summary of key findings on food consumption is provided in Box 6.1.

Box 6.1. **Residential food consumption: key findings**

Findings from descriptive analysis suggest that:

- Households' stated mean expenditure for organic fresh fruit and vegetables varies across countries and ranges from 13% to 35% of total expenditure on these products (organic and "conventional"). Environmental attitudes and norms seem to be the main determinants of expenditure on organic fruit and vegetables, and meat and poultry that take animal welfare into account. The data suggest a limited impact of socio-economic and demographic factors on mean expenditure for organic food.
- There is wide variation across countries in terms of the levels of recognition and trust in labels. For example, trust in the new EU organic food label varies from 47% in Sweden to 83% in the Netherlands among respondents who recognised it. Since there is a close link between reported expenditures and label recognition and trust, this has implications for policy makers.
- The median willingness-to-pay for organic fruit and vegetables varies from a 5% price increase in Australia and Canada, to a 23% price increase in Korea. The reported median willingness-to-pay for meat and poultry that takes animal welfare into account varies from 10% to 20%.
- Overall, respondents report that approximately 10% of food is thrown away. There is significant cross-country variation, with the median ranging from 6% in France to 14% in Israel and 15% in Korea. Younger respondents report higher levels of food waste. Those concerned with natural resource depletion are less likely to throw food away.

2. Organic food consumption

Respondents were asked to consider their overall expenditures on fresh fruits and vegetables, and provide estimates of the percentage of these expenditures that involved products labelled as organic. These percentages have to be discussed with caution though, since they are stated by the respondents and not measured using scanner-based data.

Simple statistics on mean expenditure

The percentage of households' stated expenditure for organic fruit and vegetables appears to be significantly higher than what the literature generally cites.[1] The results show variation across countries with a mean expenditure ranging between 13% in Israel and 35% in Switzerland (Figure 6.1). Expenditure shares for separate products depend to a large extent on food categories and on the country in question. In the most developed organic food markets in Europe, like Denmark's, budget shares are found to be very low for meat while the highest budget share is recorded for eggs (23%).[2] The expenditure figures in the OECD survey are probably overestimated as a consequence of the use of stated behaviour techniques, or simply as a result of the respondent's confusion of what constitutes an organic label.

Figure 6.1. **Mean percentage expenditure for organic fruit and vegetables, by country**

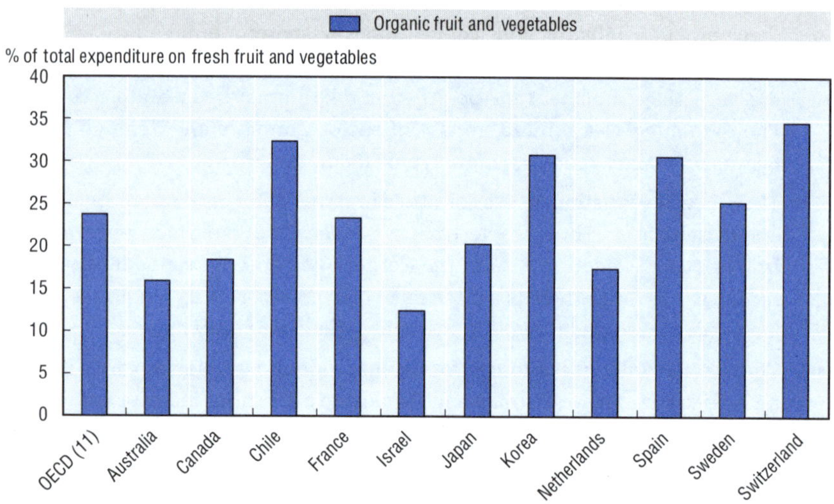

In France, the share of total consumption that was estimated to be organic – by means of a survey of stated consumption behaviour – was 24% (Agence Bio, 2009). This is in terms of physical quantities, not expenditure, but it is close to the mean expenditure share found in the 2011 EPIC Survey (23%).

Six of the eleven countries took part in the two rounds of the survey (2008 and 2011). A comparison of expenditure shares can be made with mean expenditure stated by respondents from Australia, Canada, France, Korea, the Netherlands and Sweden in 2008 (Figure 6.2)[3] with some caution as the structure of the question differed.[4] Mean expenditure calculated from the 2008 and 2011 surveys are of the same magnitude in Australia, Canada and Sweden. Reported mean expenditure is higher in 2011 than in 2008 for all

Figure 6.2. **Organic fruit and vegetables: Comparing mean expenditure in the 2008 and 2011 surveys**

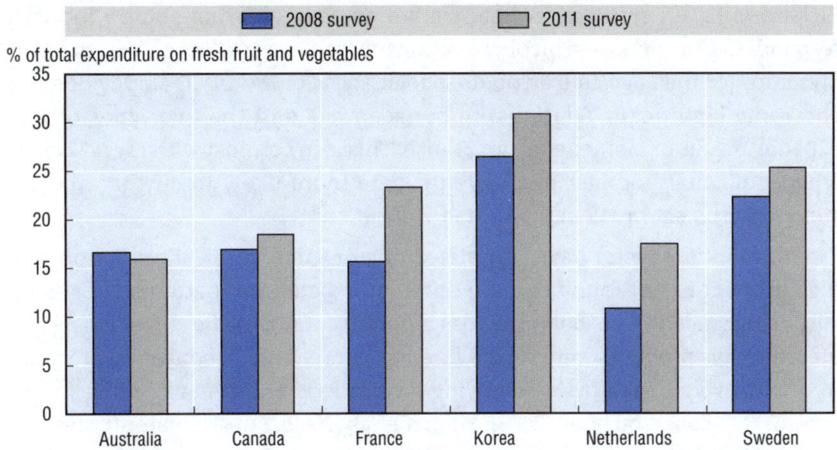

countries except Australia and the results suggest that the largest increases over that period of time have occurred in France, the Netherlands and Korea.

In the 2011 survey, respondents were also asked what percentage of their household's expenditure for meat and poultry took animal welfare into account.[5] The stated expenditure share is relatively high and varies from 23% in Japan and Korea to 53% in Switzerland and may be explained by the proliferation of private labels for meat and poultry welfare.

Statistics that may corroborate the stated expenditure shares on meat and poultry that take animal welfare into account are difficult to find. For France, Agence Bio (2009) estimates that the part of consumption that is totally *organic* is 44% for beef, 40% for poultry and 37% for other meat products (these are in terms of total consumption, not expenditure shares). In the EPIC Survey the corresponding mean for France is 34% of expenditure on meat and poultry. Respondents may indeed consider that *organic* also reflects "taking animal welfare into account".

Consumption according to respondents' characteristics

This sub-section identifies how the consumption of organic fruit and vegetables can be related to respondents' and households' socio-economic status and demographic characteristics.

Despite high price premiums for organic food, no significant pattern is found between income and expenditure neither for organic fruit and vegetables nor for meat and poultry labelled as taking animal welfare into account. The same result was obtained by Li and Zepeda (2007) and Monier et al. (2009). Some studies however found a significant relationship with income: for example, income has a positive impact in Bellows et al. (2008), and in Allender and Richards's (2010) study on animal welfare in the California poultry industry, respondents in the lower income brackets were never willing to pay the price premium for the cage-free attribute. The analysis of the OECD survey does not show any significant relationship between age and the two types of food expenditure, as in Monier et al. for France. This is in contrast with Li and Zepeda who found that younger respondents were more likely to buy organic in a survey of food shoppers in the United States.

Respondents who have completed one or more years of education after high school are not found to have different expenditure patterns. This is not surprising since we are looking across different countries and since the existing literature has produced very mixed findings. More years of education are found to be positively correlated with buying organic in Li and Zepeda (2007), Bellows et al. (2008), Epperson et al. (2008), Monier et al. (2009), but graduate studies tend to indicate a lower probability to buy organic food compared to undergraduate studies in Kidwell and Thompson (1998) and Durham (2007). There are two effects that may explain the mixed findings: further education may be linked to higher environmental awareness and hence to the purchase of organic produce; but higher education levels may also make the consumer more sceptical about the actual environmental benefits of organic farming. For example, better knowledge about agricultural production has been found to decrease willingness to pay both for organic and locally grown products (James, Rickard and Rossman, 2009). Briz and Ward (2009) confirm that higher education increases awareness of organic food and that awareness in turn increases consumption, but only at low-income and education levels.

There is no significant pattern between expenditure shares and household size and composition (presence of children under 5 years old), while the literature has found evidence of a significant relationship in some countries: the presence of children under 18 in the household positively affects the likelihood that a consumer buys an organic product in the United States (Kidwell and Thompson, 1998; Loureiro, McCluskey and Mittelhammer, 2001). The presence of children under six years of age increases the probability to buy organic milk in a study on US scanner data (Kiesel and Villas-Boas, 2007).

The absence of a relationship between having young children and mean expenditure on meat and poultry taking animal welfare into account can be observed.[6] This is not so surprising since animal welfare has more of a "public good" characteristic, while organic food has private good attributes: families with young children may increase expenditure on organic food because of food safety and health concerns. However, this could be offset by a budget effect.

Within the group of respondents who are frequently involved in purchasing decisions, women spend more on average than men for meat and poultry labelled as taking animal welfare into account. This pattern is observed in all the countries surveyed, with less marked differences in Chile, Japan and Korea. Interestingly, this is not the case for expenditure on organic fruit and vegetables. Finally, no significant expenditure pattern based on a household's location in urban (or suburban) areas versus rural areas has been observed.

The role of attitudes and values

The impacts of respondents' attitudes and opinions on food expenditure are analysed here. Environmental and health attitudes are known determinants for buying organic (Grunert and Juhl, 1995). On the one hand, Gracia and de Magistris (2008) found that Italian respondents' organic food purchasing behaviour depended more on attitudes towards the healthiness of organic produce rather than its environment-friendliness. This result is similar to those based on scanner data research revealing actual market behaviour; consumers tend to state that they value the public good aspects in a survey, but actually purchase organic food because of its private good attribute – perceived food safety and health benefit (Andersen, Millock and Wier, 2005; Griffith and Nesheim, 2008). On the other hand, Gracia and de Magistris found that the intensity of buying organic increased with environmental attitudes, but did not vary significantly with the attitudes towards the healthiness of organic produce. A similar result was found by Durham (2007) in a survey of shoppers in the Portland, Oregon, area.

Verhoef (2005) is one of the few studies that test several psychological factors on organic meat demand. In a survey of Dutch consumers, he did not find environmental *concern* to be significant in explaining stated purchase behaviour. On the other hand, *green behaviour* affected stated purchase behaviour, and perceived consumer effectiveness had a statistically significant effect both on purchase behaviour and intensity. Emotions, such as guilt, fear and empathy, affected either stated purchase behaviour or the intensity of purchase. In the Dutch study, the importance of the effects was small, though, compared to price and household socio-demographic variables. In a study of German consumers, Kuhling and Welsch (2009) also found a positive significant impact of environmental attitudes together with the behaviour of reference persons (social norms) on the consumption of organic food. The results show a positive correlation between the level of life satisfaction (measured on a 0- to 10-point scale) and mean expenditure for both organic fruit and vegetables, and meat and poultry, taking animal welfare into account. The correlation coefficient is statistically different from 0 but very small in magnitude (around 0.05).

Respondents who have supported or participated in the activities of an organisation of some sort are found to spend more on average for organic fruit and vegetables and for meat and poultry taking animal welfare into account in all countries. This includes but is not restricted to parent-teacher association, environmental organisation, local community association, charitable organisation. Israel stands as an exception for meat and poultry. When considering involvement in an environmental organisation, the same pattern is observed and so are the differences in mean expenditures between those who are involved and those who are not (Figure 6.3).

Figure 6.3. **Mean expenditure on organic fruit and vegetables and involvement in an environmental organisation**

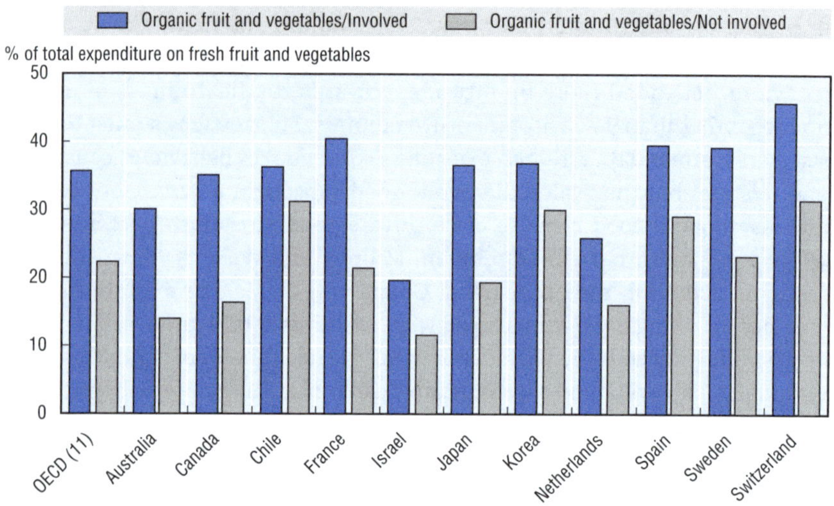

Respondents were asked to indicate what, in their view, are the most serious issues facing the world today.[7] The results suggest that, in general, mean expenditure for organic fruit and vegetables and for meat and poultry taking animal welfare into account is higher for those who view the environment as one of the most serious issues facing the world. No similar patterns can be observed between mean expenditure and the seriousness of health as a global concern. This is not really surprising for animal welfare given that it has more of a "public good" characteristic, but some relationship might have been expected for organic food because its private dimension is reflected in factors such as health benefits. Hammitt and Williams (2001), among others, show that less pesticide residue seems to be the main reason why consumers perceive organic produce as safer.

Respondents were then asked to state the degree of seriousness they attribute to a range of environmental issues[8] and, for each respondent, the index of "seriousness" was constructed.[9] Figure 6.4 shows the mean expenditure for organic fruit and vegetables separately for respondents with a "low environmental threat" perception and those with a "strong environmental threat" perception. For respondents in the "low threat" group, the index of "seriousness" is lower than the index of the median respondent in the country, while it is higher for respondents in the "strong threat" group. Both groups have an equal number of respondents.

In all countries, respondents who consider that the six environmental issues are "more serious" on average ("strong environmental threat" group) have higher mean expenditure on organic fruit and vegetables. The same patterns can be observed for mean expenditures on meat and poultry taking animal welfare into account.

Figure 6.4. **Mean expenditure for organic fruit and vegetables and seriousness of environmental issues**

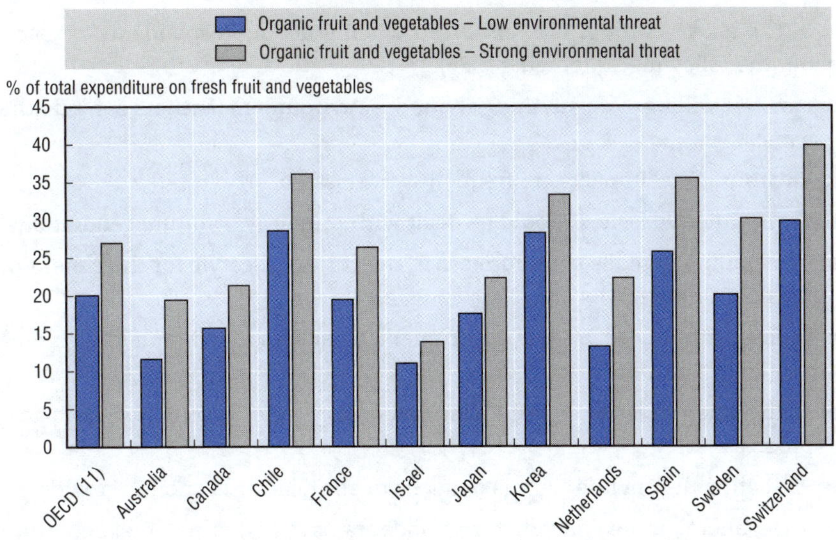

Looking more closely at the environmental concerns most directly related to consumption of organic food, one can see that the reported mean percentage of expenditures on fruit and vegetables which are organic increases with the level of reported concern about biodiversity loss (Figure 6.5). However, the relationship with concern about resource depletion is less evident, with those expressing little concern having relatively high expenditures.

Figure 6.5. **Mean expenditure for organic food and level of concern with environmental issues**

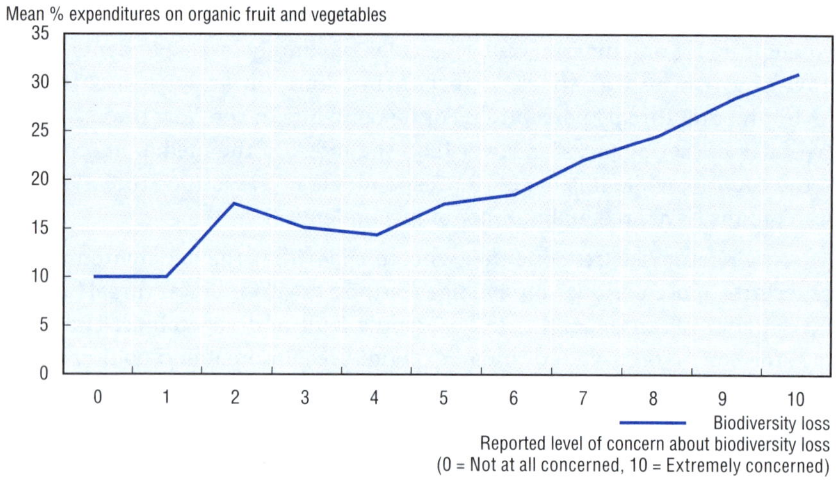

The above findings are reinforced when studying respondents' opinion about the following statements (Q26):

- I am not willing to do anything about the environment if others don't do the same.
- Environmental impacts are frequently overstated.
- Environmental issues should be dealt with primarily by future generations.
- I am willing to make compromises in my current lifestyle for the benefit of the environment.
- Policies introduced by the government to address environmental issues should not cost me extra money.
- Environmental issues will be resolved in any case through technological progress.
- Protecting the environment is a means of stimulating economic growth.

In general, more concerned respondents have higher mean expenditure than less concerned respondents, but the effect is less pronounced than is the case for the environmental index described previously. The same type of pattern is found regarding satisfaction with the local environment: those respondents who are more satisfied on average with their local environment have lower expenditure on organic fruit and vegetables. There is no significant relationship as far as expenditure for meat and poultry is concerned, though.

3. Factors that would encourage consumption of organic food

All respondents (including those who did not report that they consume organic food) were requested to indicate how important different factors were in their food shopping choices. The correlation coefficient of the expenditure on organic fruit and vegetables and ranking of these factors in food shopping choices is as follows, all of which are statistically significant:

- Personal health: 0.22
- Freshness and taste: 0.12
- Seasonal and locally produced: 0.26
- Price: -0.06
- Environmental aspects: 0.34
- Familiarity and preferred brands: 0.07

Without much surprise, those respondents ranking health, freshness and taste, seasonal and locally produced food, and environmental aspects as relatively more important factors have higher expenditures on organic fruit and vegetables. The strongest correlations are observed with "environmental aspects" and "seasonal and locally produced". Respondents ranking price as more important have lower expenditure on organic fruit and vegetables (in percentage terms).

More directly, respondents were also asked how important the following factors would be in encouraging them to increase their consumption of organic food (Q90):

- Better availability of organic products *availa*
- Lower price of organic products *price*
- Better appearance of the food *appear*
- Easier identification of organic products *identif*
- More trust in health benefits of organic products *trusthealth*
- More trust in environmental benefits of organic products *trustenvir*
- More trust in certification and labelling of organic products *trustcertif*

In Table 6.1 the factors are ordered from the most important to the least important. In all countries, lower prices would encourage respondents to buy more organic food. Among the factors that would encourage consumption of organic food, the second most important factor relates to trust: higher trust in certification and labelling (in seven countries); higher trust in health benefits (in three countries), and higher trust in environmental benefits in Korea. Trust in health benefits would more strongly encourage consumption of organic food than trust in environmental benefits in Australia, Canada, Israel, Japan,

Table 6.1. **Factors that would encourage consumption of organic food (2011 survey)**

	1 (most important)	2	3	4	5	6	7 (least important)
Australia	price	trustcertif	trusthealth	identify	trustenvir	appear	availa
Canada	price	trustcertif	trusthealth	trustenvir	identif	appear	availa
Chile	price	trustcertif	trustenvir	identify	trusthealth	availa	appear
France	price	trustcertif	trustenvir	trusthealth	identif	availa	appear
Israel	price	trusthealth	trustcertif	trustenvir	appear	identif	availa
Japan	price	trusthealth	trustenvir	trustcertif	identif	availa	appear
Korea	price	trustenvir	trustcertif	trusthealth	identif	availa	appear
Netherlands	price	trusthealth	trustenvir	trustcertif	appear	identif	availa
Spain	price	trustcertif	identif	trusthealth	trustenvir	availa	appear
Sweden	price	trustcertif	trustenvir	trusthealth	identif	appear	availa
Switzerland	price	trustcertif	trustenvir	trusthealth	identif	availa	appear

the Netherlands and Spain. In the other countries (Chile, France, Korea, Sweden and Switzerland), trust in environmental benefits is considered more important than trust in health benefits.

Availability is not considered to be very important by the respondents since it is classified as the least important factor in five countries and as the second least important factor in the other six countries. It is important to bear in mind that respondents are asked about organic food consumption in general, though. Previous studies have established supply problems for organic meat, in particular. Respondents in the survey do not seem to give much importance to this factor.

The results on the most important factors to motivate organic food consumption are very similar to the results of the 2008 OECD Household Survey. Table 6.2 shows the share of respondents who classified each factor as "very important" in 2008.

Price was the most important factor already in 2008, followed by trust in certification and trust in the health benefits provided by organic food. Availability and appearance rank last, just as in the 2011 survey. Price has been the main economic attribute studied in the literature. Some studies find that organic food buyers are less sensitive to price than conventional buyers (Hammitt and Williams, 2000). Scanner data studies of actual purchases find high price elasticities, implying that a 10% increase in price causes a more than 10% decrease in demand (Glaser and Thompson, 2000; Smed and Wier, 2000). On the other hand, a policy experiment in the Netherlands showed very low responsiveness to a reduction in the price of organic food (Bunte et al. 2010), mainly because shoppers who do not buy organic were not well informed about the reduction in price.

Table 6.2. **Factors that would encourage consumption of organic food (2008 survey)**

%

	Better availability of organic products	Lower price of organic products	Better appearance of the food	More trust in health benefits of organic products	More trust in environmental benefits of organic products	More trust in certification and labelling of organic products
Australia	21	55	19	29	28	38
Canada	23	58	20	34	31	42
France	19	68	18	31	31	42
Korea	7	36	7	35	22	18
Netherlands	12	48	16	19	18	22
Sweden	23	55	24	32	34	39

This echoes the result for France in Monier et al. (2009) that shows that marginal price decreases have no effect on the decision to buy organic rather than conventional products, whereas the demand for eggs among organic food consumers was price-elastic as in the Danish and US studies. However, the demand for organic milk was relatively price-inelastic. Organic food is also a substitute for conventionally produced food (Glaser and Thompson, 2000),[10] and sometimes for other eco-labelled food. Loureiro, McCluskey and Mittelhammer (2001) found that eco-labelled produce and organic produce are competing for consumers with similar attitudes towards food safety and that this type of consumer would prefer to buy organic at equal prices.

Respondents in the 2011 survey were also asked how useful information on animal welfare would be on a product label (on a scale from 0: not useful, to 10: very useful). Table 6.3 reports the average score in each country.

Table 6.3. **Usefulness of information on animal welfare**

	Average score (0 to 10)
Australia	7.1
Canada	7.3
Chile	8.2
France	7.0
Israel	7.1
Japan	6.0
Korea	6.1
Netherlands	6.8
Spain	7.4
Sweden	7.4
Switzerland	7.9

No significant correlation was found between the above scores and demographics (age, income). The countries where the information is considered the most useful, on average, are also those with the highest average willingness-to-pay for meat and poultry that take animal welfare into account (Chile and Switzerland).

4. Food waste, food "miles" and animal welfare

Respondents were asked to state whether they agree or disagree with the following statements (see Table 6.4):

a) Consuming meat and other animal products has significant negative environmental consequences.

b) Importing food from distant areas has significant negative environmental consequences.

c) Food waste has significant negative environmental consequences.

The answers have been recoded so that a negative score indicates disagreement (a maximum of 2) and a positive score agreement (a maximum of +2).

Table 6.4. **Opinions on relationship between food consumption and environment**

	Consuming meat, etc.	Importing food	Food waste	Awareness index
Australia	-0.36	0.65	0.56	0.28
Canada	-0.29	0.54	0.61	0.29
Chile	-0.43	-0.04	0.18	-0.10
France	-0.20	1.09	1.19	0.69
Israel	-0.26	0.11	0.72	0.19
Japan	0.03	0.55	0.46	0.35
Korea	0.13	0.21	1.05	0.47
Netherlands	-0.23	0.49	0.89	0.38
Spain	-0.60	0.29	0.10	-0.07
Sweden	-0.18	1.12	1.28	0.74
Switzerland	-0.32	0.96	0.96	0.53

Apart from Korea and Japan, respondents in all countries do not agree that meat consumption has negative environmental consequences. With the exception of Chile (where there is a neutral stance), respondents in all countries agree that importing food from distant areas has significant negative environmental consequences. On average, there is also slight agreement with the statement that food waste has negative environmental consequences. A positive, but not very strong, correlation (0.11-0.13) can be observed between the aggregated "awareness index" and expenditure on both organic food and meat and poultry that take animal welfare into account.

Food waste can represent a high proportion of household waste and raises particular concerns in some countries like Israel and Korea. In order to provide some insights on this issue, in the 2011 survey, respondents were also asked to estimate the percentage of food bought by their household which was thrown away. While in most of the countries surveyed the median value[11] is 10%, the results presented in Table 6.5 show significant variations with the median ranging from 6% in France to up to 14% in Israel and 15% in Korea.

Table 6.5. **Median percentage of food that is thrown away**

	Median (%)
Australia	10
Canada	10
Chile	11
France	6
Israel	14
Japan	10
Korea	15
Netherlands	10
Spain	10
Sweden	10
Switzerland	7

Environmental concerns are not found to have a significant impact on the proportion of food thrown away by households. Also, there is no significant relationship between food waste and a household's decision to compost its food waste.

The results however suggest that young generations tend to waste more food, as Figure 6.6 illustrates. There is also a positive and significant correlation between food waste and household size, which may reflect the difficulty of planning food consumption over time as household size increases. However, the correlation is generally quite small and is only statistically significant in six of eleven countries.

In general, there is a positive correlation between food waste and expenditure on organic fruit and vegetables. The largest correlation coefficient is for Korea (0.34). In only one case (Switzerland) is the correlation statistically insignificant (see Table 6.6).

Is the generation of food waste related to concern for the environment? In Figure 6.7 the mean percentage food waste is compared with the rank given to environmental issues relative to five other public concerns. Although there

Figure 6.6. **Food waste by age group, OECD(11)**

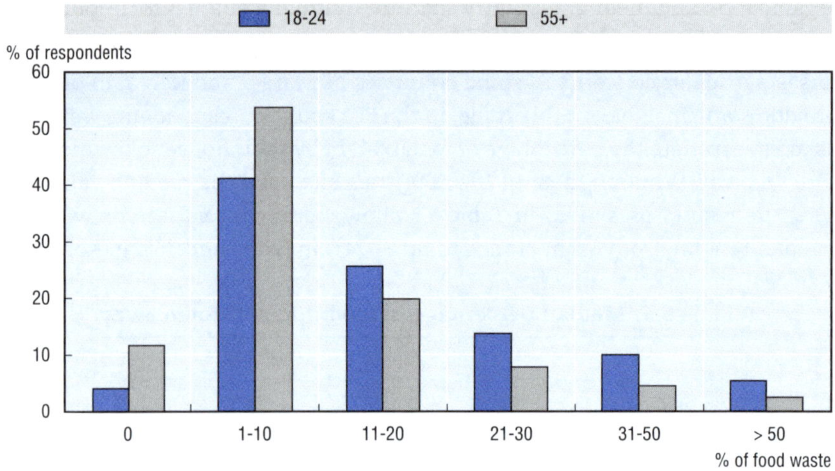

Table 6.6. **Food waste and organic food consumption**

	Correlation coefficient between food waste and expenditure on organic fruit and vegetables
Australia	0.23
Canada	0.18
Chile	0.24
France	0.14
Israel	0.25
Japan	0.17
Korea	0.34
Netherlands	0.14
Spain	0.25
Sweden	0.09
Switzerland	0.05

Note: All significant at 5% level, except for Switzerland.

is a slight upward trend, with those ranking environmental issues relatively low (fourth to sixth) having higher levels of food waste, the differences are not highly significant.

Figure 6.8 compares food waste generation with reported concern for solid waste in general, as well as with natural resource depletion. The only apparent relationship relates to those with a low level of concern for natural resource depletion having significantly higher rates of food waste.

Figure 6.7. **Percentage of food that is thrown away and rank of environmental concerns**

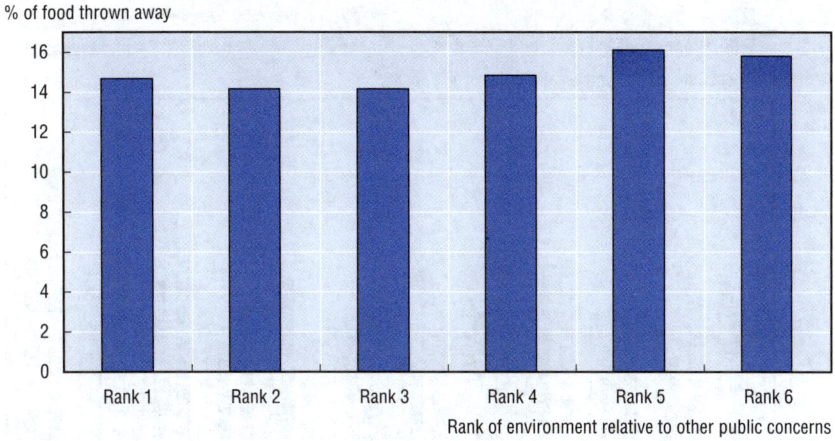

Figure 6.8. **Percentage of food that is thrown away and stated concern for selected environmental issues**

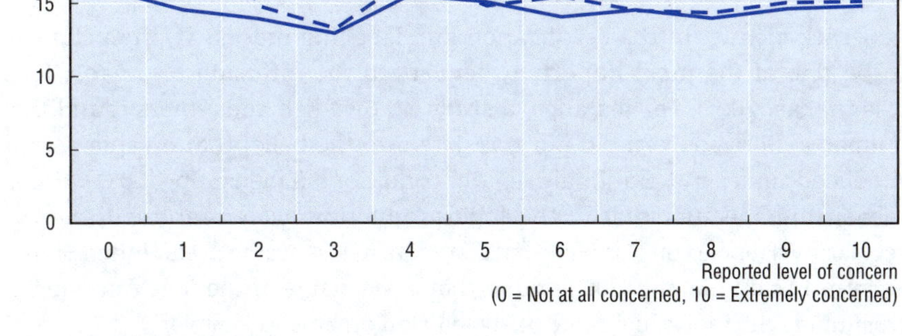

5. Environmental labelling and trust

Respondents were asked how trustworthy they consider a number of sources with regard to information on claims about the environmental impacts of products. In general, respondents indicating higher trust also have higher mean expenditure. Figure 6.9 shows separately the mean percentage expenditure for meat and poultry and organic fruit and vegetables for two groups of respondents: those in the "low trust" group have lower trust in

Figure 6.9. **Mean percentage expenditure and trust in manufacturers and retailers**

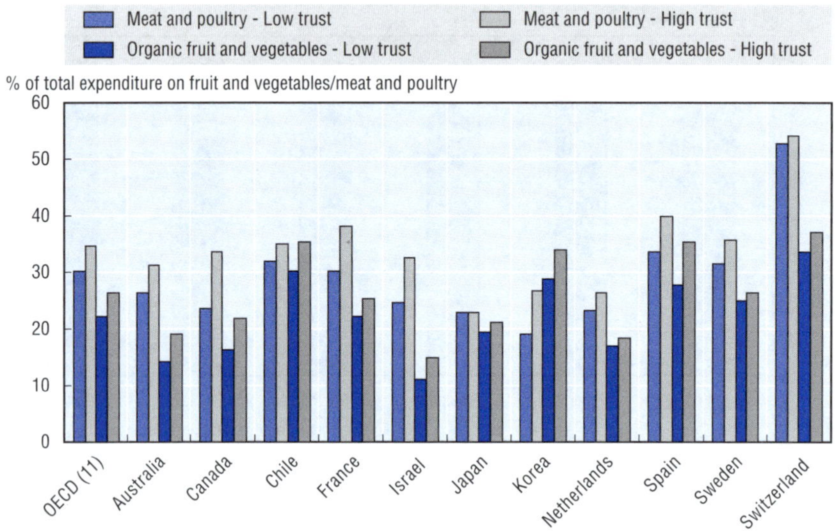

manufacturers and retailers than the median respondent in the country, while those in the "high trust" group trust manufacturers and retailers more than the median respondent. In all countries respondents who consider manufacturers and retailers as "more" trustworthy have higher mean expenditure for organic fruit and vegetables as well as for meat and poultry taking animal welfare into account.

Lack of trust in the certification and labelling process is often claimed to be one of the most important barriers to the diffusion of organic food (Giannakas, 2002). For instance, a study on Swedish consumers found that animal welfare concern did not translate into the belief that organic pork is produced under more animal-friendly conditions (Liljenstolpe, 2011). When measuring specific trust in the food production system and trust in the regulatory agencies on a sample of consumers in the Boston area, United States, Hammitt and Williams (2000) found that a low value of the index measuring trust made the individual more likely to be an organic food buyer.

Australia is the only country in which two different national logos for organic food have been displayed. The European Union (EU) food labels (old and new) have been shown to respondents from the four European countries: France, the Netherlands, Spain and Sweden. The Netherlands and Switzerland are the only countries in which there is an official logo for food produced according to animal welfare principles. Hence, the forthcoming analysis is based on the set of logos that have been displayed at the time the survey took place. Some other organic food labels exist in most countries but have not been presented on-screen to the respondents.

In France and Sweden, more than 90% of the respondents recognise the national organic food label (Table 6.7). This percentage is also quite high in Switzerland (82%). On the contrary, in Australia, Canada, Israel, Japan and Spain, less than 30% of the respondents recognise the national organic food logo.

Table 6.7. **Organic food and animal welfare labels**

	Definition	Logo is recognised	For those who recognised the logo		
			Label is understood	Label is trusted	Label is used
Australia	National organic label (Australian Certified Organic)	28%	96%	81%	52%
Australia	National organic label (NASAA Certified Organic)	11%	89%	83%	61%
Canada	National organic label	29%	96%	79%	51%
Chile	National organic label	32%	96%	92%	70%
France	National organic label	93%	99%	78%	53%
France	EU old organic label	32%	91%	77%	53%
France	EU new organic label	19%	56%	65%	38%
Israel	National organic label	29%	88%	73%	38%
Japan	National organic label	24%	66%	75%	43%
Korea	National organic label	60%	99%	74%	76%
Netherlands	National organic label	64%	84%	85%	46%
Netherlands	Animal welfare label (Beter Leven)	38%	96%	90%	60%
Netherlands	EU old organic label	13%	96%	76%	53%
Netherlands	EU new organic label	8%	55%	83%	41%
Spain	National organic label	28%	96%	88%	66%
Spain	EU old organic label	32%	97%	88%	65%
Spain	EU new organic label	10%	60%	73%	48%
Sweden	National organic label (KRAV)	97%	96%	80%	54%
Sweden	EU old organic label	23%	92%	78%	40%
Sweden	EU new organic label	13%	45%	47%	16%
Switzerland[1]	Nationwide organic label	82%	97%	80%	50%

1. Note that the BioSuisse label is not a government-sponsored label, and that many other organic food labels are used in Switzerland. For example, many grocery store chains in that country use their own organic label.

In the four European countries (France, the Netherlands, Spain and Sweden), the old EU organic food label (that was used on a voluntary basis) is better known than the new one, that was made compulsory for all pre-packaged organic food products on 1 July 2010. It had been in use only for little more than half a year when the survey was undertaken. Indeed, maybe for reasons of novelty, just as fewer respondents recognise and understand the new label, it is also less trusted than the old one on average. The only exception occurs in the Netherlands, where respondents are more likely to trust the new label than the old one (although fewer respondents state that they understand it). It should also be noted that more respondents trust the new label than claim that they understand it. Incidentally, this is also the case for the Japanese national organic label.

In France, the Netherlands and Sweden, respondents are found to be much more aware of the national organic food labels than of the EU ones. In all countries except Japan, between 84% and 99% of the respondents who recognised the national label were also able to understand it. In general, there are also many respondents who recognised a label and who also trusted it. On average across all countries, about half of the respondents who recognised a logo also used it.

To corroborate the figures on recognition of labels, Agence Bio (2009) estimates that 87% of the French recognise the label AB, and that 39% recognise the old European label. Among consumers of organic food, recognition of the label is 98% for the AB label and 58% for the European label. The figures in the OECD 2011 survey were 93% and 32% of the French sample, respectively.

In some of the countries that have the lowest percentage recognition of labels, consumers may be confused by a multitude of different competing labels. In Australia, for instance, there are as many as six different labels for organic food and the market share of each label varies across the different states.

As in the previous round of the OECD survey, Sweden and France have the highest level of recognition of the national label (97% and 87% respectively in 2008). The level of recognition of the Canadian national organic label has increased to 29% compared to 18% in the 2008 survey (OECD, 2011) while the figures seem to be stable for Australia.

Some simple statistics show that older respondents (55+) are more likely to recognise the national organic label than younger respondents (18 to 34 age group). Respondents aged 35 to 44 are the most likely to use the label in their purchasing decisions. Wealthier respondents are more likely to consider the organic food label in their purchasing decisions, compared to poorer respondents. No correlation was found between the use of organic labels and respondents' level of education.

Dutch respondents were the only group in the survey to whom their country's animal welfare label was shown and questioned about.[12] In the Netherlands, respondents who recognise the national animal welfare label estimate at 34% their household's food expenditures for meat and poultry labelled as taking animal welfare into account, while the average expenditure share is 18% for respondents who do not recognise the national label. The same magnitude in expenditure shares is observed between respondents who trust the national label (37%) and those who do not (18%).

As far as organic fruit and vegetables are concerned, respondents who recognised the national organic label and those who trust it have higher mean expenditures (Figure 6.10).

Figure 6.10. **Mean percentage expenditure on fruit and vegetables labelled as organic and national organic label recognition and trust**

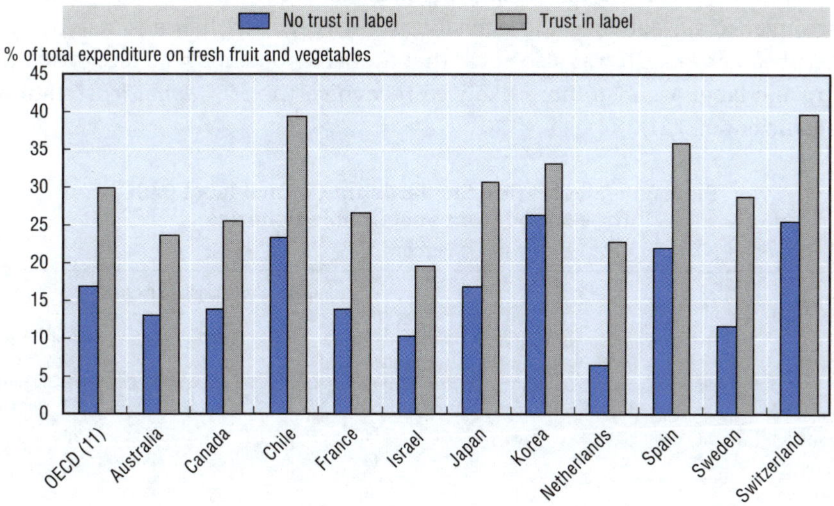

Table 6.8 shows that a similar pattern exists for the European old and new labels.

Table 6.8. **Mean percentage expenditure and European label recognition and trust**

%

	No label recognition	Label recognition	No trust in label	Trust in label
France	19	29	23	32
Netherlands	15	26	18	30
Spain	29	33	24	37
Sweden	22	33	26	37

These findings confirm those found in the literature and in particular that households buying organic food are generally more knowledgeable about organic foods (Chang et al., 2006) and are more likely than conventional buyers to rate organic food labels as important (Briz and Ward, 2009; Hammitt and Williams 2000). Some studies show that the coexistence of several different labels confuses consumers (Abrams et al., 2010); in the United States claims as to "all-natural" products were seen as less trustworthy and induced some confusion with the organic label.

Respondents from all countries except Chile and Korea were also surveyed about recognition, understanding, trust and use of the Marine Stewardship Council's (MSC) label for seafood from sustainable fisheries (Table 6.9). Those respondents who recognise the MSC label are more likely to choose fish certified as sustainable over other types of fish (the correlation coefficient of 0.22 is highly significant). As expected, the correlation is stronger (0.49) between the decision to buy certified fish and use of the certified fish label. It was also noted that those respondents who recognise the organic label are also more likely to recognise the MSC label (correlation coefficient of 0.21).

Table 6.9. **The Marine Stewardship Council label (MSC) for seafood from sustainable fisheries**

Percent

	Logo is recognised	For those who recognised the logo		
		Label is understood	Label is trusted	Label is used
Australia	6	91	76	59
Canada	8	96	92	72
France	3	73	68	52
Israel	4	69	85	43
Japan	6	86	86	70
Netherlands	13	89	82	65
Spain	10	86	85	67
Sweden	20	91	83	62
Switzerland	30	93	88	71

MSC reports that its own commissioned marketing survey on consumers in the United Kingdom, the United States, Germany, Japan, Canada and France found that 23% of the adult population recognise the MSC label, compared to 8% in 2008.

6. Willingness-to-pay (WTP) for organic produce and animal welfare

Respondents were asked to state the maximum percentage price increase they would be willing to pay for organic food compared to a conventional substitute. The question was asked only for one food category[13] "fresh fruit and vegetables" as findings in the 2008 round of the survey indicated that willingness-to-pay was consistent across food groups in the countries surveyed. In the 2011 survey, a similar question was also introduced to analyse the stated willingness-to-pay for meat and poultry labelled as taking animal welfare into account.

Level of willingness-to-pay

The results suggest variations in WTP for fresh fruit and vegetables among countries. Figure 6.11 shows that the median WTP[14] varies from a 5% price increase in Australia and Canada, to a 23% price increase in Korea. The mean WTP varies from 16% in Australia and the Netherlands to 34% in Korea.

Figure 6.11. **Median willingness-to-pay for fresh fruit and vegetables labelled as organic**

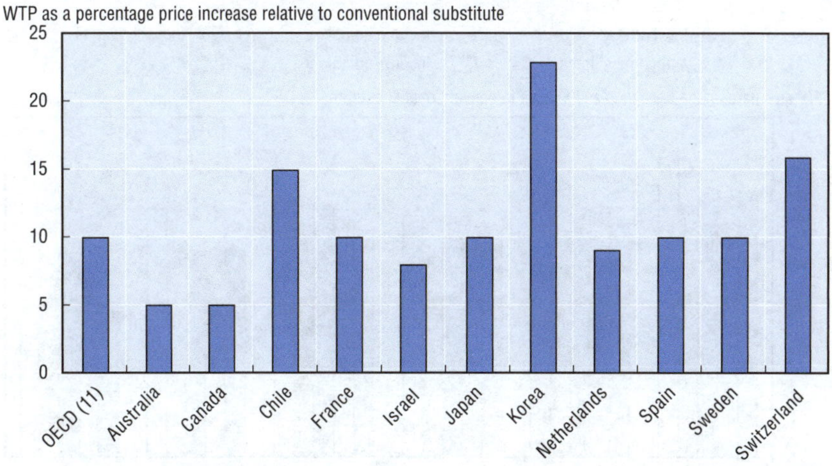

For the six countries where both rounds of the survey were implemented (Australia, Canada, France, Korea, the Netherlands and Sweden), some form of comparison can be made with the 2008 results, bearing in mind that the structure of the response format differed in the two rounds.[15]

The share of respondents who are not willing to pay any extra money for organic food is more or less the same between 2008 and 2011, except in the Netherlands where there are fewer respondents not willing to pay extra in 2011. In all countries, respondents are willing to pay more, on average, in 2011 than in 2008 but, again, this may be explained partly by the way answers could be formulated.

Willingness-to-pay and households' characteristics

The results show no general pattern between income and WTP, except that respondents in the highest income quintile report having a higher WTP for meat and poultry taking animal welfare into account in all countries except Chile (the second quintile has the highest WTP), Korea (first quintile) and Sweden (second quintile). The wealthiest households state a higher WTP for organic fruit and vegetables in six countries: Canada, Israel, Japan, Korea, the Netherlands and Switzerland.

Figure 6.12 shows some pattern in terms of WTP for different income and age groups for the sample as a whole. Older respondents are willing to pay less in general for animal welfare and organic produce than younger respondents (as in Govindasamy and Italia, 1999). Among each age group, respondents in the fifth quintile have the highest WTP.

Figure 6.12. **Mean willingness-to-pay by age group and income quintile, OECD(11)**

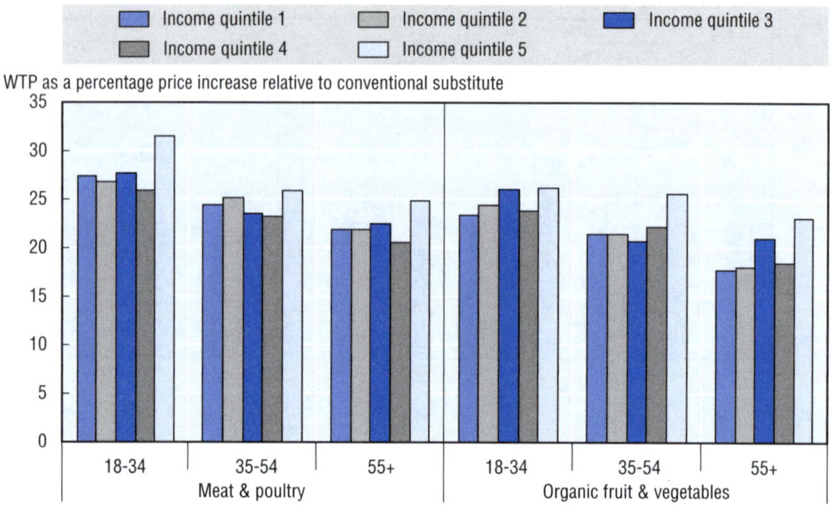

Respondents with at least one year of post high school education have a higher WTP in general, with the exception of Chile and Korea for both meat and poultry and organic fruit and vegetables. As was the case for mean expenditure, there is no significant relationship between household size and composition (presence of young children), and WTP. Within the group of respondents who are frequently involved in purchasing decisions, women state a higher WTP than men for meat and poultry labelled as taking animal welfare into account (28% versus 21% for men) and for organic fruit and vegetables (23% versus 20% for men). On a country-by-country basis, respondents living in urban or suburban areas are found to have a higher WTP for organic fruit and vegetables in general.

The role of attitudes

Life satisfaction and WTP are positively correlated but the correlation coefficient is rather small (0.05). While there seems to be no significant relationship between stated WTP and the respondent's decision to vote in an election, WTP is positively correlated with community involvement, and in particular supporting or participating in the activities of an environmental organisation. This is the case for both animal welfare and organic produce (Figure 6.13)

Figure 6.13. **Mean willingness-to-pay and involvement in an environmental organisation**

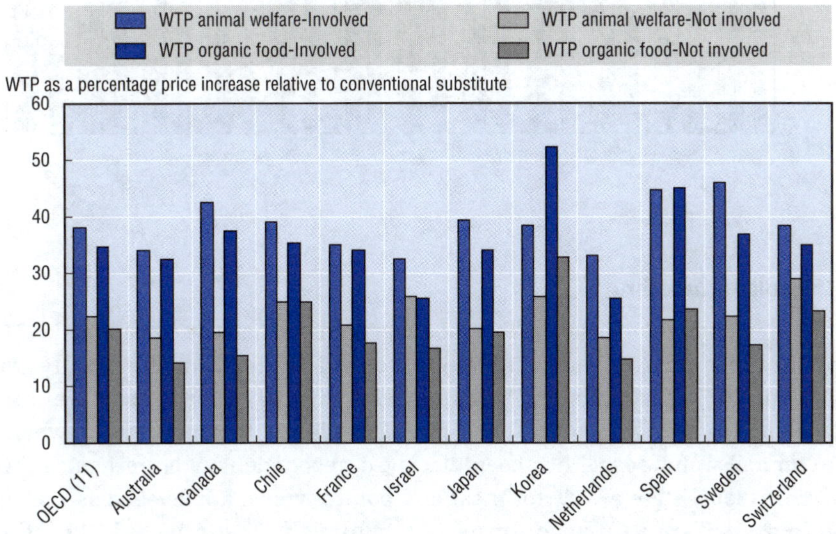

A similar pattern exists between WTP and seriousness of concern about the environment, as that which was observed for mean expenditures. In general respondents who are more concerned about the environment or who consider environmental issues as more serious state a higher WTP. There also seems to be some pattern between WTP and satisfaction with the local environment: respondents who are more satisfied with their local environment have a lower WTP in general. As for expenditures, a positive, but not very strong, correlation (0.13-0.16) is observed between the aggregated "awareness index" and average WTP for both organic food and meat and poultry that take animal welfare into account. In addition, a positive relationship between trust (in manufacturers and retailers) and WTP seems to hold, as was the case for mean expenditure (Figure 6.14).

Figure 6.14. **Mean willingness-to-pay and trust in manufacturers and retailers**

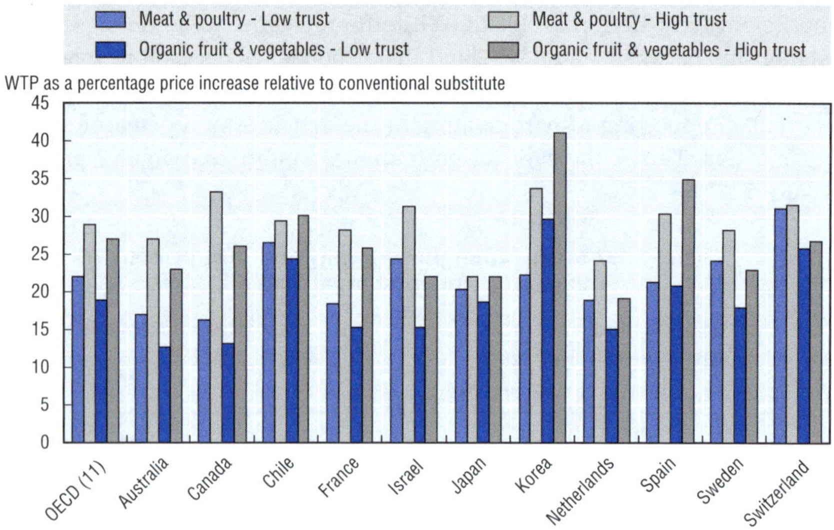

The role of labelling

Previous work has found that labelling tends to increase consumers' WTP for organic milk (Kiesel and Villas-Boas, 2007). In another study on the United States, consumers displayed a high WTP for animal welfare labelling (Tonsor and Wolf, 2011). In the Netherlands, the only country where an animal welfare label has been displayed, respondents who recognise the national label state a WTP for meat and poultry which are labelled as taking animal welfare into account at 27%, while the mean WTP is 17% for respondents who do not recognise the national label. The same magnitude in WTP is observed between respondents who trust the national label (30%) and those who do not (12%).

As far as organic fruit and vegetables are concerned, it is found that respondents who recognised the label and those who trust the label have a higher WTP in general (Figure 6.15). In France, Korea and Sweden, those who recognised the national organic label have a lower WTP in general. However, these figures have to be interpreted with caution because the percentage of respondents who did not recognise the label is very small in France (7%) and in Sweden (3%). There are thus only a few observations in the group of respondents who did not recognise the national organic label.

The same pattern is observed for the European organic labels (both old and new ones), see Table 6.10.

Figure 6.15. **Mean willingness-to-pay and label recognition and trust**

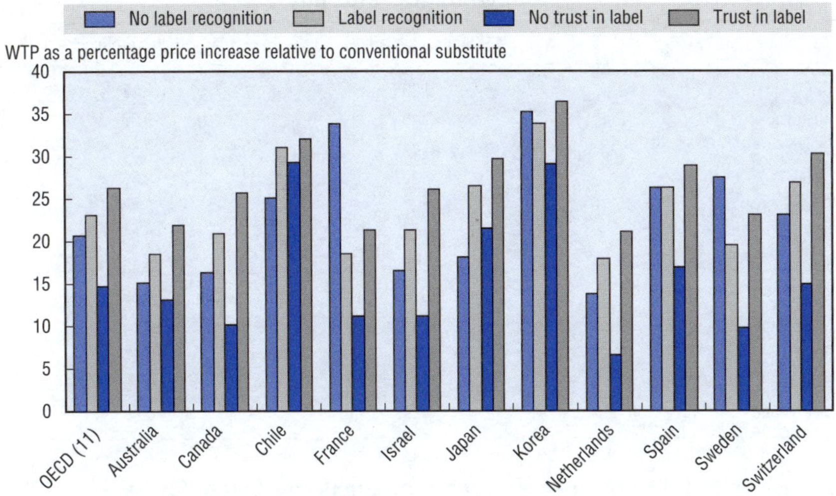

Table 6.10. **Mean willingness-to-pay and European label recognition and trust**

Percent

	No label recognition	Label recognition	No trust in label	Trust in label
France	18	21	16	23
Netherlands	16	20	13	24
Spain	24	31	26	33
Sweden	16	28	18	32

Link between expenditure and willingness-to-pay

There is a significant positive correlation between expenditure and WTP in each country, for both organic fruit and vegetables and animal welfare (Table 6.11). This is not really surprising *per se* in the sense that the respondents who buy this type of products are, by definition, willing to pay the premium for food that takes animal welfare into account and for organic produce.

Reasons for not willing to pay extra

Those respondents who declared that they were not willing to pay extra for organic fresh fruit and vegetables were asked to state why this was the case. For each of the four answers listed below, respondents had to state whether they strongly disagree (-2), disagree (-1), agree (1), strongly agree (2) or have no opinion (0).

- Answer 1: If I had more money I would pay more for organic food
- Answer 2: I do not trust the food is actually grown organically

Table 6.11. **Correlation coefficients between expenditure and willingness-to-pay**

	Animal welfare	Organic produce
OECD (11)	0.39	0.52
Australia	0.40	0.57
Canada	0.47	0.59
Chile	0.32	0.36
France	0.37	0.46
Israel	0.41	0.66
Japan	0.52	0.50
Korea	0.56	0.55
Netherlands	0.51	0.60
Spain	0.34	0.46
Sweden	0.43	0.48
Switzerland	0.24	0.48

- Answer 3: I do not believe organic products are better for health or the environment
- Answer 4: I do not think I should have to pay extra

The corresponding index, which is the simple average of the opinions in each country, is shown in Table 6.12. In all countries except Korea, most respondents are not willing to pay extra because "they do not think they should have to pay extra" (answer 4). In Korea, the main reason for not wanting to accept any price increase is that respondents "do not trust the food is actually grown organically". For respondents in all countries except Chile and Japan, money is not the primary reason for not wanting to pay extra (the index corresponding to answer 1 is negative).

Table 6.12. **Reasons for not wanting to pay extra**

	If I had more money I would pay more for organic food (1)	I do not trust the food is actually grown organically (2)	I do not believe organic products are better for health or the environment (3)	I do not think I should have to pay extra (4)
Australia	-0.37	0.39	0.25	1.30
Canada	-0.48	0.35	0.24	1.22
Chile	0.27	0.10	-0.76	1.18
Israel	-0.33	0.40	0.46	1.39
Japan	0.03	0.11	0.00	1.05
Korea	-0.08	1.29	-0.06	1.16
Netherlands	-0.48	0.52	0.32	1.32
Spain	-0.22	0.53	-0.28	1.29
Sweden	-0.43	0.61	0.11	1.34

7. Conclusions

The analysis focused on how socio-economic factors, attitudes and labelling influence the expenditure and willingness to pay (WTP) for organic fruit and vegetables with some insights on meat and poultry labelled to take animal welfare into account. The descriptive statistics and simple correlations presented point to some policy issues to be further examined by means of appropriate econometric techniques.

Factors explaining expenditure

Households' stated mean expenditures for organic fresh fruit and vegetables vary across countries and range from 13% to 35%. These figures are higher than those reported in the literature and may be an overestimation by households, as often is the case in surveys of stated behaviour, or simply the result of the respondent's confusion of what constitutes an organic label.

Results suggest a limited impact of socio-economic and demographic factors on mean expenditure for organic food. There is no significant correlation between mean expenditure and income, age, education, household size or residence.

Attitudes and behaviour seem to be the main determinants of expenditure on organic fruit and vegetables, and meat and poultry that take animal welfare into account. Previous research shows that it is the perceived private health benefits rather than the environmental benefits that affect actual purchase behaviour. Respondents who are involved in an organisation of some sort (and in particular an environmental association) are found to have higher mean expenditures than respondents who have not supported or participated in such organisations.

Attitudes such as the perceived seriousness of an environmental threat are found to increase the average mean expenditure. In all countries, respondents who consider the listed environmental issues as more serious have higher mean expenditure on average for both organic fruit and vegetables and meat and poultry taking animal welfare into account. In addition, respondents who are more satisfied with the local environment have lower expenditure on organic fruit and vegetables.

Trust has been shown in previous studies to be a major barrier against buying organic food. The OECD survey asked respondents to indicate the trustworthiness of a number of sources with regard to information on claims about the environmental impact of a product. Respondents who expressed higher trust also had a higher percentage of their expenditures on organic fruit and vegetables and meat and poultry taking animal welfare into account. In particular, respondents who consider manufacturers and retailers as more trustworthy have higher mean expenditures on both meat and poultry and organic fruit and vegetables than respondents with low trust in manufacturers and retailers. Increasing trust in the labelling of organic products appears to be key. Advertising the health and environmental benefits of organic food is a possibility, but this has to be supported by scientific evidence. Restoring trust could be a difficult task, in particular after the identification of an organic farm as the likely source of the *E. coli* outbreak of foodborne illness that occurred in Germany in May-June 2011. Increasing trust in certification and labelling seems to be an important policy objective, in particular in Korea. The increasing number of (private and public) labels in relation to agricultural practices and food quality may have induced some confusion among consumers.

The main factors that would encourage further consumption of organic food have not changed since the 2008 OECD Household Survey: it is still a lower price, and more trust in the certification and labelling of organic products comes second. This result is compatible with the stated factors driving food choice, primarily "freshness and taste", "price" and "health". Hence, consumers may be willing to buy organic food as long as they believe or trust that these products are fresh and tasty, as well as good for their health, and as long as they can afford the price.

Affordability of organic food is found to be the most important concern for households in the lowest income quintile and for respondents aged 18 to 24. The fact that organic food is perceived as "too expensive" may not only reflect a problem of affordability. Indeed, there seems to be, in a number of cases, some misunderstanding or lack of knowledge. Some respondents do not seem to understand why organic food is more expensive than conventional food. In the follow-up questions on willingness to pay for organic food, the most common reaction in all countries, except Korea, is "I don't think I should have to pay extra". This may reflect a need for improved communication between the agricultural sector and the general public who does not seem to be aware of the higher costs induced by organic farming compared to conventional farming. It may also be a result of the more widespread distribution of organic products in standard food chain outlets and a concern that part of the price premium does not correspond to the higher cost of production. Differences in perception of organic produce could be further explored in more formal econometric analysis.

Food waste and miles

The responses to the survey indicate that only in Korea and Japan do people generally agree with the statement that meat consumption has negative environmental consequences. However, a majority of respondents in all countries agree that importing food from distant areas has significant negative environmental consequences. On average, there is also slight agreement with the statement that food waste has negative environmental consequences.

The results however suggest that young generations tend to waste more food. There is also a positive and significant correlation between food waste and household size, which could reflect the difficulty of planning food consumption over time as household size increases. However, the correlation is generally quite small and is only statistically significant in six of eleven countries. Also, no significant relationship is found between food waste and a household's decision to compost its food waste.

In general, there is a positive correlation between food waste and expenditure on organic fruit and vegetables. Comparing food waste with the rank given to environmental issues relative to five other public concerns, it is found that those ranking environmental issues relatively low (i.e. fourth to sixth) have higher levels of food waste. With respect to specific environmental issues, the only apparent relationship relates to those with a low level of concern for natural resource depletion having significantly higher rates of food waste generation.

The role of labelling

The results on labelling differ according to the country and the label considered (national organic label, EU organic label). A number of labels (public or private) may coexist in the countries studied, but the analysis presented here is based on a subset of labels that was shown to the respondents. In some of the countries that have the lowest recognition of labels, consumers may be confused by a multitude of different competing labels. In the case of Australia, for example, there are in fact six different labels for organic food (the market share of each label varies across states).

Results show significant variation across countries in terms of label identification: on the one hand, in France and Sweden more than 90% of the respondents recognise the national organic food label. Label recognition is also quite high in Switzerland (82%). On the other hand, the national organic food logo is recognised by less than 30% of the respondents in Australia, Canada, Israel, Japan and Spain. The old EU label is better known than the new one in the four European countries, but it is difficult to draw conclusions since the new one had been in use only for little more than half a year when the survey was undertaken.

On average across all countries, about half of the respondents who recognised a logo also used it. Older respondents (55+) are more likely to recognise the national organic label than younger respondents (18-34). Those aged 35 to 44 are the most likely to use the label in their purchasing decisions. Wealthier respondents are more likely to consider the organic food label in their purchasing decisions, compared to poorer respondents. In order to use a label, consumers also need to trust its claims. Indeed, respondents who recognise the national organic label (or the European organic label) and those who trust it have higher mean expenditure than those who do not recognise the label and do not trust it. A general policy conclusion, despite the differences between countries, is that there is scope for increased awareness and knowledge of organic food labels, in particular among young respondents (18 to 34).

Willingness-to-pay

The median WTP for meat and poultry that take animal welfare into account varies from 10% to 20% (in Switzerland) while the mean WTP varies from 20% in Australia to 31% in Switzerland. There is much more variation in the WTP for organic fresh fruit and vegetables across countries. The median WTP for fresh organic fruit and vegetables varies from a 5% price increase in Australia and Canada, to a 23% price increase in Korea. The corresponding mean WTP varies from 16% in Australia and the Netherlands to 34% in Korea.

Older respondents are willing to pay less in general for organic produce and animal welfare than younger respondents. Among each age group, respondents in the highest income quintile have the highest WTP. Community involvement, in particular involvement in an environmental organisation, is positively correlated with WTP for both organic produce and animal welfare. Trends between WTP and the perceived seriousness or concern about the environment are similar to those observed for mean expenditure. Higher trust also seems to indicate higher WTP for both types of products.

Respondents who recognise the label and who trust it also have a higher WTP in general. In the Netherlands, the only country for which an animal welfare label was displayed, respondents who recognise and trust the label stated a higher WTP than those who do not recognise it and do not trust it. The results for WTP for both animal welfare and organic fruit and vegetables are thus strikingly similar to the factors that determine stated expenditure.

However, in order to draw firm policy recommendations, the findings of the descriptive analysis need to be further examined using multivariate econometric analysis. This work is ongoing.

Notes

1. Bunte et al. (2010) state an overall budget share on organic food of 2.3% in 2009 in the Netherlands and 3.4% in Switzerland. In Sweden, the total expenditure share on organic food was 3.4% in 2008 and 4.0% in 2009 according to Statistics Sweden. Expenditure shares for separate products vary to a large extent depending on product. According to Smed and Wier (2000), for fresh produce, the organic budget share was 4% in the Netherlands in 1997-2000, compared with 17% in the survey. Part of this difference is certainly attributable to the lapse in time.

2. Smed and Wier (2000) find that Denmark had very low organic meat budget shares over the 1997-2000 years (5.8% for lamb and 2.2% for minced meat) and the highest budget share for organic milk and eggs (23%).
3. Note that the set of respondents is not the same in 2008 and in 2011. Hence the comparison between the two surveys will be meaningful only if the sample of respondents in each country is representative of the population in each country. As noted in Introduction and Annex A, the sample is close to being representative.
4. The 2011 questionnaire used a slider while in the 2008 questionnaire answers on expenditure for organic fruit and vegetables were in the form of intervals for all values except 0% and 100%: [1% to 5%], [6% to 10%], [11% to 25%], [26% to 50%], [51% to 75%] and [76% to 99%]. The mid-point of each interval is considered in order to calculate the mean expenditure in each of the six countries.
5. No direct questions were asked on meat and poultry labelled to take animal welfare into account in the 2008 OECD questionnaire.
6. The absence of relationship remains for both types of food expenditure, even when controlling for household size.
7. Respondents had to rank the following six issues from 1 (the most important) to 6 (the least important) (Q22): *a)* International tensions; *b)* Economic concerns; *c)* Environmental concerns; *d)* Health concerns; *e)* Social issues; and *f)* Personal safety.
8. Respondents had to state the degree of seriousness they attribute to the following environmental issues on a scale from 0 to 10 (Q23): *a)* Waste generation; *b)* Air pollution; *c)* Climate change (global warming); *d)* Water pollution; *e)* Natural resource depletion (forest, water, energy); and *f)* Endangered species and biodiversity.
9. The index was constructed by calculating the simple average of the scores given to the six issues in Q23. A higher index indicates that the respondent considers the listed environmental problems are "more serious".
10. In economic terms, they are close substitutes, with high cross-price elasticities.
11. The median value is reported rather than the mean since there are some outliers in the responses.
12. Animal welfare labels are also available in other countries, for example in Switzerland, but these labels were not known by the researchers at the time the 2011 survey was administered.
13. The 2008 OECD survey included other food categories such as "eggs" and "milk and dairy products".
14. Calculation includes those reporting a willingness-to-pay equal to zero.
15. A slider was used in the 2011 survey with a continuous scale from 0% to 200% (or more) while intervals were presented to respondents in the 2008 survey: 0%, [1% to 5%], [6% to 15%], [16% to 30%], [31% to 50%], [>50%].

References

Abrams, K.M., C.A. Meyers and T.A. Irani (2010), "Naturally Confused: Consumers' Perceptions of All-Natural and Organic Pork Products", *Agriculture and Human Values*,Vol. 27, No. 3, pp. 365-374.

Agence Bio (2009), *Baromètre de consommation et de perception des produits biologiques en France*, Rapport n° 0901164 – Edition 2009.

Allender, W.J. and T.J. Richards (2010), "Consumer Impact of Animal Welfare Regulation in the California Poultry Industry", *Journal of Agricultural and Resource Economics*,Vol. 35, No. 3, pp. 424-442.

Andersen, L.M., K. Millock and M. Wier (2005), "Information Provision, Consumer Perceptions and Values – the Case of Organic Foods", In Krarup, S. and C. Russell (Eds.) *Environment, Information and Consumer Behaviour*, Edward Elgar New Horizons in Environmental Economics, pp. 161-178.

Bellows, A.C. et al. (2008), "Understanding Consumer Interest in Organics: Production Values vs. Purchasing Behavior", *Journal of Agriculture and Food Industrial Organization*,Vol. 6, No. 1, Article 2.

Boccaletti, S. (2008), "Environmentally Responsible Food Choice", *OECD Journal: General Papers*, Vol. 2008, No. 2, pp. 117-152.

Briz, T. and R.W. Ward (2009), "Consumer Awareness of Organic Products in Spain: An Application of Multinomial Logit Models", *Food Policy*,Vol. 34, No. 3, pp. 295-304.

Bunte, F.H.J. et al. (2010), "Limits to Growth in Organic Sales", *De Economist*,Vol. 158, No. 4, pp. 387-410.

Chang, H.S., C. Leviten-Reid and L. Zepeda (2006), "Organic Food Demand: A Focus Group Study Involving Caucasian and African-American Shoppers", *Agriculture and Human Values*, Vol. 23, No. 3, pp. 385-394.

Durham, C. (2007), "The Impact of Environmental and Health Motivations on the Organic Share of Produce Purchases", *Agricultural and Resource Economics Review*,Vol. 36, No. 2, pp. 304-320.

Epperson, J.E. et al. (2008), "Modeling Fresh Organic Produce Consumption with Scanner Data: A Generalised Double Hurdle Model Approach", *Agribusiness*,Vol. 24, No. 4, pp. 510-522.

Giannakas, K. (2002), "Information Asymmetries and Consumption Decisions in Organic Food Product Markets", *Canadian Journal of Agricultural Economics*,Vol. 50, No. 1, pp. 35-50.

Glaser, L.K. and G. Thompson (2000), *Demand for Organic and Conventional Beverage Milk*, Selected paper presented at the Western Agricultural Economics Association Annual Meeting, Vancouver, British Columbia, 29 June-1 July.

Govindasamy, R. and J. Italia (1999), "Predicting Willingness to Pay a Premium for Organically Grown Fresh Produce", *Journal of Food Distribution Research*, Vol. 30, No. 2, pp. 44-53.

Gracia, A. and T. de Magistris (2008), "The Demand for Organic Foods in the South of Italy: A Discrete Choice Model", *Food Policy*,Vol. 33, pp. 386-396.

Griffith, R. and L. Nesheim (2008), *Household Willingness to Pay for Organic Products*, CEPR Discussion Paper 6905, CEPR.

Grunert, S.C. and H.J. Juhl (1995), "Values, Environmental Attitudes, and Buying of Organic Foods", *Journal of Economic Psychology*,Vol. 16, pp. 39-62.

Hammitt, J.K. and P.R.D. Williams (2000), "A Comparison of Organic and Conventional Fresh Produce Buyers in the Boston Area", *Risk Analysis*,Vol. 20, No. 5, pp. 735-746.

Hammitt, J.K. and P.R.D. Williams (2001), "Perceived Risks of Conventional and Organic Produce: Pesticides, Pathogens, and Natural Toxins", *Risk Analysis*,Vol. 21, No. 2, pp. 319-330.

James, J.S., B.J. Rickard and W.J. Rossman (2009), "Product Differentiation and Market Segmentation in Applesauce: Using a Choice Experiment to Assess the Value of Organic, Local and Nutritional Attributes", *Agricultural and Resource Economics Review*,Vol. 38, No. 3, pp. 357-370.

Kidwell, J. and G. Thompson (1998), "Explaining the Choice of Organic Produce: Cosmetic Defects, Prices and Consumer Preferences", *American Journal of Agricultural Economics*,Vol. 80, No. 2, pp. 277-287.

Kiesel, K. and S. Villas-Boas (2007), "Got Organic Milk? Consumer Valuations of Milk Labels after the Implementation of the USDA Organic Seal", *Journal of Agricultural and Food Industrial Organization*,Vol. 5, pp. 1-38.

Kuhling, J. and H. Welsch (2009), "Determinants of Pro-Environmental Consumption: The Role of Reference Groups and Routine Behavior", *Ecological Economics*,Vol. 69, No. 1, pp. 166-176.

Li, J. and L. Zepeda (2007), "Characteristics of Organic Food Shoppers", *Journal of Agricultural and Applied Economics*,Vol. 39, No. 1, pp. 17-28.

Liljenstolpe, C. (2011), "Demand for Value-Added Pork in Sweden: A Latent Class Model Approach", *Agribusiness*,Vol. 27, No. 2, pp. 129-146.

Loureiro, M. L., J. McCluskey and R. C. Mittelhammer (2001), "Assessing Consumer Preferences for Organic, Ecolabeled and Regular Apples", *Journal of Agricultural and Resource Economics*, Vol. 26, No. 2, pp. 404-416.

Monier, S. et al. (2009), "Organic Food Consumption Patterns", *Journal of Agricultural and Food Industrial Organization*, Vol. 7, No. 2, pp. 12.

OECD (2011), *Greening Household Behaviour: The Role of Public Policy*, OECD Publishing. doi: *10.1787/9789264096875-en*.

Smed, S. and M. Wier (2000), *Forbrug af økologiske fødevarer, Del 2: Modellering af efterspørgseln* ("Consumption of Organic Foods, Part 2", In Danish, with English summary), Technical Report No. 319.

Thompson, G. (1998), "Consumer Demand for Organic Foods: What We Know and What We Need to Know", *American Journal of Agricultural Economics*,Vol. 80, No. 5, pp. 1113-1118.

Tonsor, G.I. and C. Wolf (2011), "On Mandatory Labelling of Animal Welfare Attributes", *Food Policy*,Vol. 36, No. 3, pp. 430-437.

Verhoef, P.C. (2005), "Explaining Purchases of Organic Meat by Dutch Consumers", *European Review of Agricultural Economics*,Vol. 32, No. 2, pp. 245-267.

Greening Household Behaviour
Overview from the 2011 Survey

Chapter 7

Household waste generation, recycling and prevention

by
Ofira Ayalon, Sharon Brody *and* Mordechai Shechter*

This chapter presents an overview of the survey data on the determinants of households' waste-related behaviour and examines the impact of waste charges and recycling programmes on waste generation and separation rates and waste prevention efforts. The role of general attitudes towards the environment in influencing household behaviour is considered as well.

* The University of Haifa, Israel.

The statistical data for Israel are supplied by and under the responsibility of the relevant Israeli authorities. The use of such data by the OECD is without prejudice to the status of the Golan Heights, East Jerusalem and Israeli settlements in the West Bank under the terms of international law.

1. Introduction

Changes in lifestyle, the increasing use of disposable materials and excessive packaging are all contributing to the amount of waste being generated. Waste management is now not only a local, but rather a global concern, affecting land and water quality, but also climate change.

Municipal waste encompasses all the wastes produced by households, offices and retail, but excludes industrial, hazardous and construction wastes. Between 1980 and 2005, municipal waste generation within the OECD countries has increased by as much as 2.5% per year, while the increase in the number of OECD households has only risen by 0.8% per year. Assuming that no new policies are introduced, the total municipal waste generation is likely to increase by a further 1.3% per year until 2030 (OECD, 2008). According to OECD statistics for countries with available data, household waste usually represents over half of the municipal waste stream.

Current waste policies throughout the OECD aim at diverting increasing amounts of valuable materials from landfills to recycling and reuse. Nevertheless, the continuously increasing environmental pressures from waste streams are forcing governments to reassess their waste management policies. This includes implementing more initiatives, such as extended producer responsibility programmes and improved collection schemes, to encourage households and businesses to recycle more of the waste they generate.

This descriptive analysis focuses on how household characteristics – including their general attitude towards the environment – are associated with their waste disposal and recycling behaviours. The key questions addressed here are:

- How do the general attitudes to the environment correlate with waste generation levels, recycling levels and waste prevention?
- Do unit-based waste fees have significant effects on waste generation or waste recycling relative to flat (or no) fees?
- To what extent is households' waste generation affected by attributes of waste-related public services, such as frequency of waste collection?
- To what extent do households' waste recycling decisions depend upon the attributes of recycling programmes (door-to-door collection, drop-off)?
- Does labelling of consumer goods with their recycled content influence purchasing decisions?

- What affects preferences for alternative government actions to reduce household waste generation?
- How do households across the countries surveyed dispose of some types of hazardous waste, such as e-waste or pharmaceutical products?

This chapter addresses these questions through a descriptive analysis of the data gathered thanks to the 2011 Survey on Environmental Policy and Individual Behaviour Change (EPIC) The following sections successively present the peer-reviewed research on policy impacts and the role of household characteristics in reducing waste; the data on waste generation, recycling, and separation; attitudes towards different waste policies and how they relate to households' own decisions about waste generation and recycling. The chapter concludes by highlighting lessons one can draw from the data to improve the efficiency of waste policies.

A summary of key findings on waste generation, recycling and prevention is provided in Box 7.1.

Box 7.1. **Waste generation, recycling and prevention: key findings**

Findings from descriptive analysis suggest that:

- Waste generation tends to be between 20% and 30% lower with unit pricing by volume or weight. Moreover, where there is unit pricing there is an increase in separation. However, relatively few respondents reported that they were subject to unit pricing, except in Korea, Japan and Switzerland.
- The presence of recyclable material collection services reduces waste generation rates and increases recyclable waste separation rates. The relative effect of door-to-door vs. drop-off collection services on waste separation rates is less apparent. This was not the case in the 2008 survey, indicating the need for further research.
- Awareness of service availability plays an important part in household separation and recycling. Overall, 19% of households report that they do not know what collection services are available for recyclables in their area. This may result in more recyclable materials being discarded with mixed waste.
- Household hazardous waste such as electronic components and (particularly) old or unused medicines are commonly disposed of inappropriately. In the latter case 34% is disposed of with general mixed waste. Rates are particularly high for younger people, over 50% for many countries.
- The two policy measures which respondents supported in terms of waste generation rates relate to waste prevention – namely to encourage retailers to use less packaging and households to purchase products with less packaging.

2. Research on the impacts of waste policies and the role of households' characteristics

Various studies analyse how socio-economic parameters affect households' waste generation. One particularly subtle relationship is that between household income and waste generation/disposal. A fairly robust body of evidence has illustrated that higher-income households generate more waste (Afroz et al., 2010; Bandara et al., 2007; Adams et al., 1993; Jenkins, 1993; Jenkins et al., 2003; Mazzanti and Zoboli, 2009). However, some dissenting studies find no relationship between waste generation and income (Karbassi et al., 2012; Badruddin et al., 2002). Furthermore, higher-income households in many settings tend to recycle and separate their waste more frequently (Callan and Thomas, 1997; Duggal et al., 1991; Ferrara and Missios, 2005; Hong, 1999; Jenkins et al., 2003).

Household size has also been highlighted as a key characteristic determining waste generation: Studies have also shown that, while larger households naturally produce more waste (Bandara et al., 2007; Mazzanti and Zoboli, 2009; Fujiwara et al., 2010), the waste generated per person is usually lower in larger households (Afroz et al., 2010). Age is also highly correlated with the amount of waste separated and recycled, with middle-aged and older people more likely to recycle (Meneses and Palacio, 2005; Nixon et al., 2006; Ebreo and Vining, 1990). Further, research also suggests that older people more often comply with social norms (Bruvoll and Nyborg, 2004; Berglund et al., 2009).

Education likely plays a key role across countries in households' decisions on how to manage their waste. Higher education has been found to be associated with lower waste generation (Karbassi et al., 2012) and also with an increase in separation and recycling (Callan and Thomas, 1997; Duggal et al., 1991; Ferrara and Missios, 2005; Jenkins et al., 2003; Reschovsky and Stone, 1994).

The collection system available in a given community clearly plays an important role in determining separation rates. Door-to-door recycling programmes are estimated to increase separation rates by up to 100% compared to drop-off centres (Ashenmiller, 2011; Best, 2009; Dahlén and Lagerkvist, 2010; Dahlén et al., 2007; Fullerton and Kinnaman, 2000; Reschovsky and Stone, 1994).

Various incentive schemes have also been implemented throughout OECD countries, in order to make households internalise the costs of – and thereby reduce – their waste generation. For example, pay-as-you-throw (PAYT) schemes have been piloted in many municipalities throughout OECD countries. Not surprisingly, the effectiveness of PAYT programmes appears to depend on their characteristics, and on attributes of the communities in which they are implemented. Weight-based billing for waste disposal generally decreases waste generation by around 20%, but leaves the proportion of recycled waste essentially unchanged (Callan and Thomas, 2006; Åberg et al., 2009).

Studies indicate that combining door-to-door collection with PAYT billing can increase the sorted waste by an extra 12% although there is no reduction in total waste (Allers and Hoeben, 2010; Bucciol et al., 2011; Hong, 1999; Blume et al., 1994; Joshi et al., 2010). One drawback with the introduction of PAYT billing that has been observed in some contexts is an increase in illegal dumping (Fullerton and Kinnaman, 1994).

Reward and incentive schemes such as bring back/refund schemes also contribute to motivation regarding waste reduction and recycling (Tam and Tam, 2007). Ashenmiller (2011) found that women, married people, older age groups, and low-income families were more responsive to deposit-refund and other cash reward systems.

3. Waste generation

The role of socio-demographic factors

In the questionnaire, respondents were asked how many bags of waste they put out for disposal on average per week. To help respondents estimate their waste generation in terms of number of bags, a visual depiction of the bag sizes were used (Figure 7.1). Although this response format is an improvement over the 2008 EPIC survey (described in OECD, 2011), there are still some aspects of waste generation that the data do not measure. For example, households generating the same volume or weight may differ greatly in the weight of generated waste. Indeed, volume-based pricing can induce changes in the density of waste because households pack their bags more tightly.[1]

Figure 7.1. **Image displayed to respondents to help them estimate their household waste**

Nevertheless, data on volume-based waste generation are useful as a simple indicator for comparing waste generation rates and their determinants across countries. Using this metric (Figure 7.2), it can thus be seen for instance that households in Korea evidently generate the least amount of mixed waste per capita (13 litres on average per week) compared to ten other countries (42 litres on average). Israeli respondents, on the other hand, appear to be the most intensive producers of household waste out of any of the total sample, with 69 litres generated per week per capita.

In terms of how waste generation relates to households' demographic and socio-economic characteristics, the most obvious factor to consider is household size. As mentioned above, previous research found that waste generation increases with household size, but often at a decreasing rate. The 2011 EPIC data indicate that indeed waste generation increases with household size, but that the rate of increase varies significantly from one country to the next. Most countries – with the exception of Australia, the Netherlands and Sweden – exhibit "economies of scale" at the household level with regard to waste generation: in eight of the eleven countries, waste generation per person decreases with every additional person in the household.

Results from the data are mixed on the question of whether waste generation rises with income (Figure 7.3). No such relationship is evident for four countries: Canada, Chile, the Netherlands and Spain. And, although a positive correlation can be seen for the other seven countries, even here the relationship is not clear, with waste generation sometimes falling from one

Figure 7.2. **Average weekly household waste per person, by country and household size**

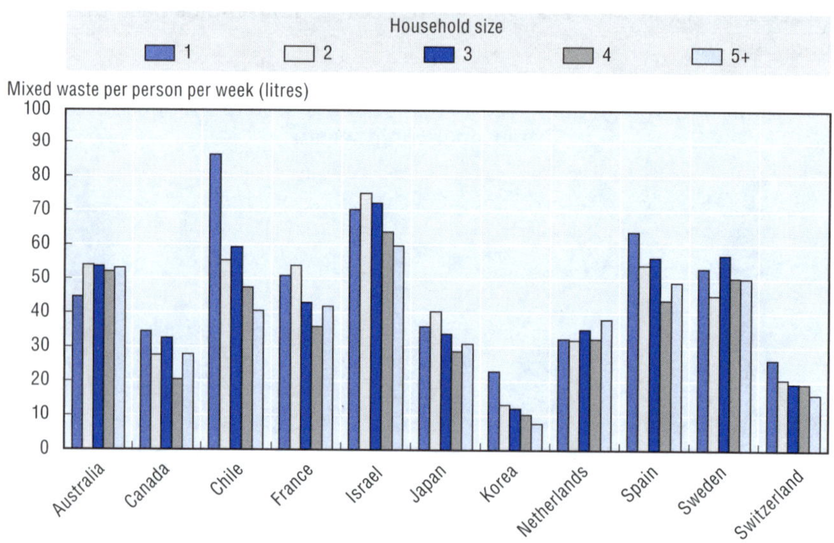

Figure 7.3. **Average weekly household waste, by country and by income quintile**

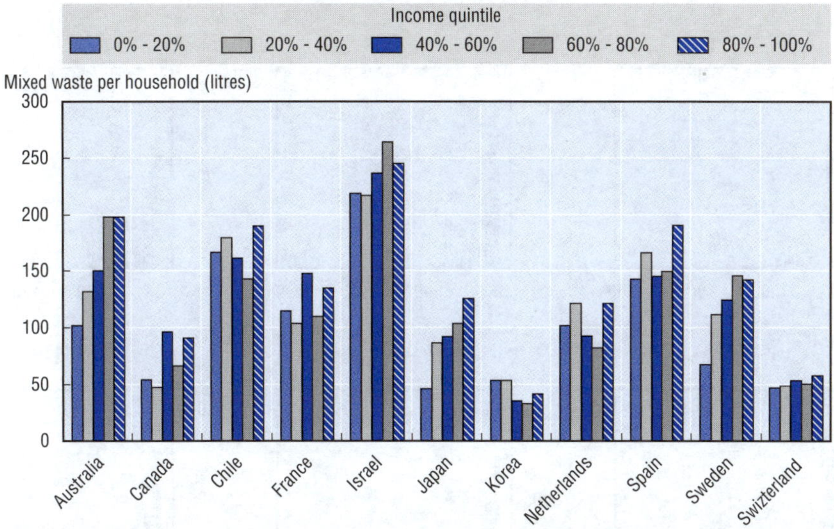

income quintile to the next, even while the general trend is positive. In general, the data evidently support the mixed findings from previous research on the relationship between waste generation and income. Whether or not the correlation is positive appears to depend on country-specific characteristics.

Other household characteristics were investigated but were not found to be significantly correlated with waste generation when controlling for other related factors. These include education, the size of the current residence and the length of time residing in it. However, attitudinal variables appear to have an influence. Figure 7.4 presents the relationship between waste generation rates and the rank of environmental concerns relative to other public policy objectives (such as security, health, among others). The distribution of reported waste generation volumes is lower for those who rank environmental issues higher.

Figure 7.5 presents the relationship between an index of environmental attitudes and waste generation rates. The index was derived from responses to a set of seven questions related to the environment.[2] Those respondents whose index is relatively low generate considerably more mixed waste. However, this effect is less pronounced as one moves up the index. Indeed, for values greater than the mid-point of the index (0), the effect is not apparent.

The role of waste charging schemes

As mentioned in the literature reviewed above, the billing system for municipal waste disposal services should have a strong impact on the amount of household waste generated. In particular, it is more economically efficient

Figure 7.4. **Average weekly household waste by rank of environmental issues**

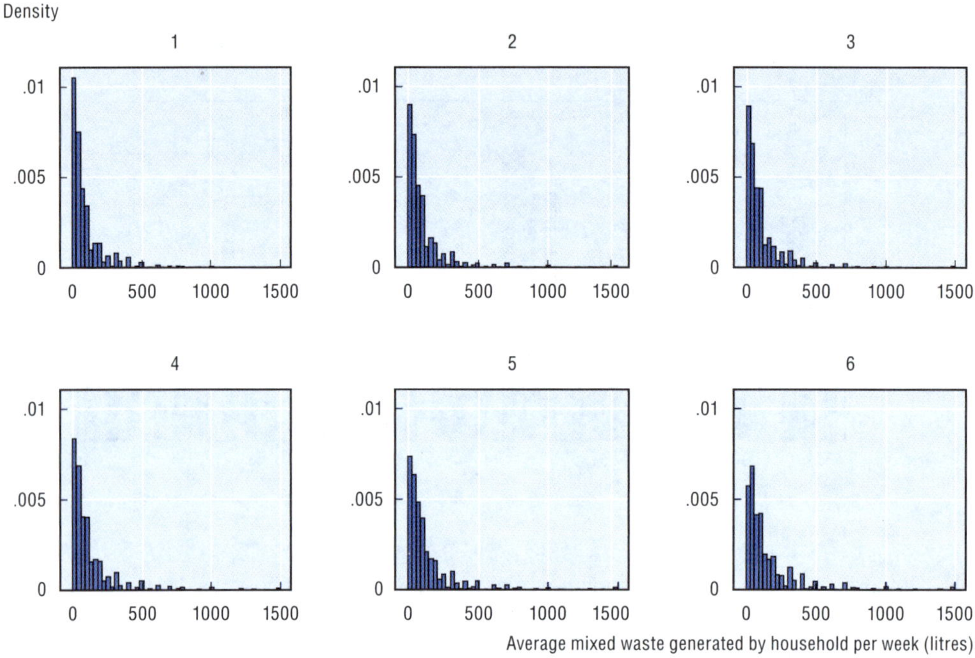

Note: Rank 1 stands for the most serious issue and 6 for the least important.

Figure 7.5. **Average weekly reported waste generated and index of environmental attitudes**

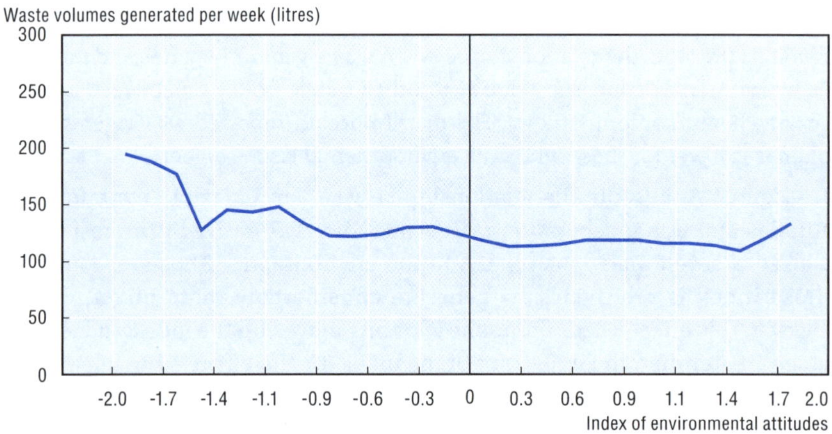

– and in keeping with the Polluter Pays Principle – for households to face a unit-based charge for their mixed waste. As indicated in Figure 7.6, there is still a high percentage of households who report that they do not pay for their waste according to the amounts generated. Some report paying a flat fee (e.g. included in property charges), while others say that they are not charged at all. As noted in Chapter 1, it is likely that a small percentage of respondents did not respond accurately to questions such as this.

Figure 7.6. **Households reporting having a pay-as-you-throw and other billing systems for mixed waste disposal, by country**

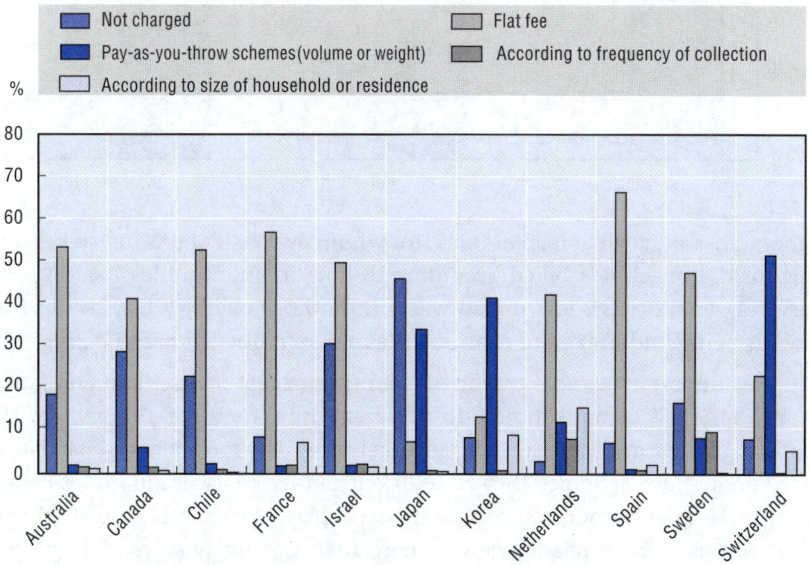

However, in some countries unit-based systems – often referred to as pay-as-you-throw (PAYT) billing systems – are relatively common. These include weight-based and volume-based charges. PAYT implementation is the highest in Switzerland (53%), Korea (42%) and Japan (35%). Households in countries using such systems are found to dispose of less mixed waste than those charged a simple flat rate (between 25% and 33% as Figure 7.7 suggests). The formal analysis of the data will allow to cast further light on these results. In the rest of the countries, the prevalence of PAYT is too low to reliably estimate the impact on household waste generation. The percentage of respondents reporting having a pay-as-you-throw system for mixed waste disposal is the lowest in Spain, where a flat fee is by far the dominant billing mode (67%) as illustrated in Figure 7.6.

The evidence with respect to frequency-based charging is ambiguous. In the Netherlands, where frequency-based billing appears to be relatively common, there does not seem to be any statistically significant effect on waste generation. However, on the basis of the survey data on waste

Figure 7.7. **Comparison of volume-based versus flat fee impacts on waste generation in Japan, Korea and Switzerland**

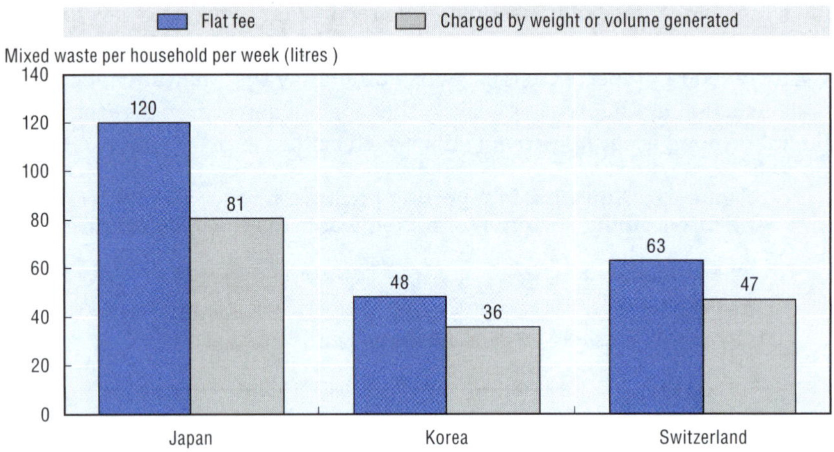

generation for Swedish respondents, for whom over half of PAYT systems take this form, households billed according to how frequently they use disposal services generate 23% less mixed waste than those paying a flat fee. Further work is being undertaken to examine the role of frequency-based billing.

It is important to note that in all countries except Switzerland and Japan, more than 10% of households do not know how they are charged for the disposal of their mixed waste. It is very likely that they too are charged a flat fee since under such schemes they face little incentive to ascertain how they are charged. However, other factors may also be at play – for example the number of years the household has resided in its residence is likely a predictor of how knowledgeable its members are of their waste billing scheme (Figure 7.8).

Figure 7.8. **Respondents not knowing how they are billed for their mixed waste according to length of residence**

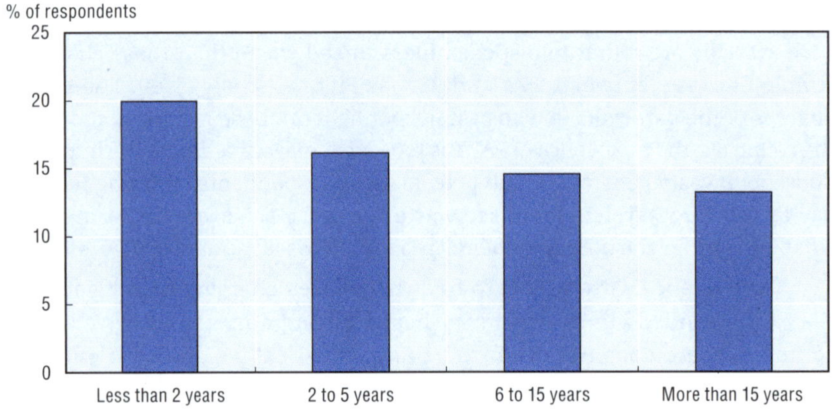

4. Disposal of waste containing hazardous materials

Although much of the study has concentrated on mixed waste generation and sorting by households, data on the disposal of waste containing hazardous materials have also been examined. Two common types of wastes causing potential environmental and health damage were considered: old and unused medicines and old electronic equipment.[3] Medicines contain a wide range of chemicals and old electronic equipment may contain high levels of heavy metals such as lead and cadmium. Figure 7.9 presents the data on whether or not respondents dispose of these wastes as part of the household mixed waste collection.

Figure 7.9. **Respondents disposing of their hazardous waste with their mixed waste collection**

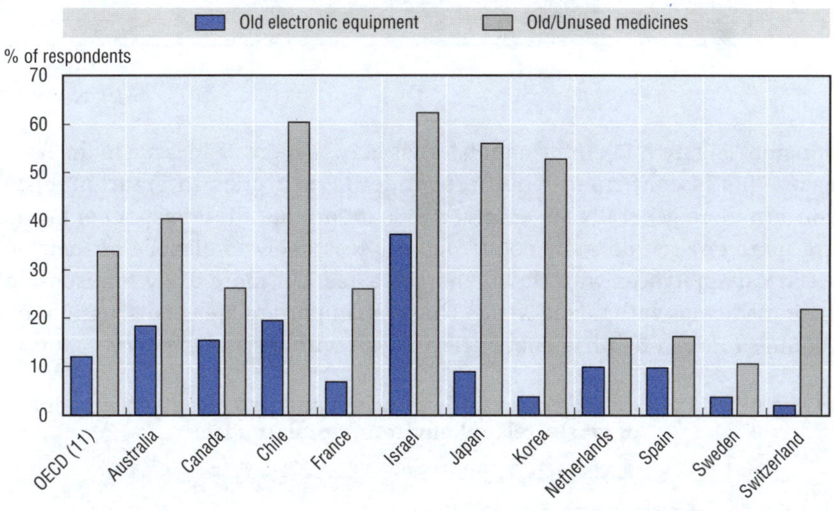

Across the countries, 12% of respondents reported disposing of their old electric equipment with their mixed waste collection, and 34% disposed of old or unused medicines in their mixed waste. Moreover, whether or not respondents disposed of medicines improperly was highly correlated with whether or not they disposed of old electronics in a similar manner. As Figure 7.10 shows, older respondents are more likely to dispose of both waste streams separately, whereas the younger age groups are more likely to dispose of them with the mixed waste for collection.

The relationship between respondents' assessment of the seriousness of waste generation as an environmental issue and the likelihood that they dispose of medicines and electronics with mixed waste is not particularly strong. While those who are more concerned about waste tend to be less likely to dispose of electronics with mixed waste, in the case of medicines this is less

Figure 7.10. **Households disposing of their hazardous waste with mixed waste collection, by age group**

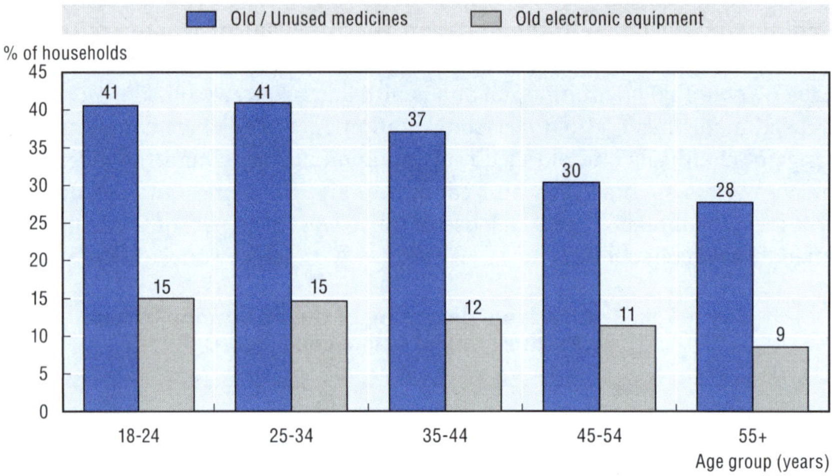

apparent (Figure 7.11). Information issues may be more of an issue in the latter case. This is confirmed when respondents' attitudes to environmental concern more generally are assessed. It is found that those respondents who are more environmentally concerned are less likely to dispose of their old electrical appliances with their general wastes, but more likely to dispose of their medicines with mixed waste. This is an important area of discussion that requires more in-depth econometric analysis with respect to policy making.

Figure 7.11. **Disposing of hazardous waste and perceived seriousness of waste-related environmental impacts**

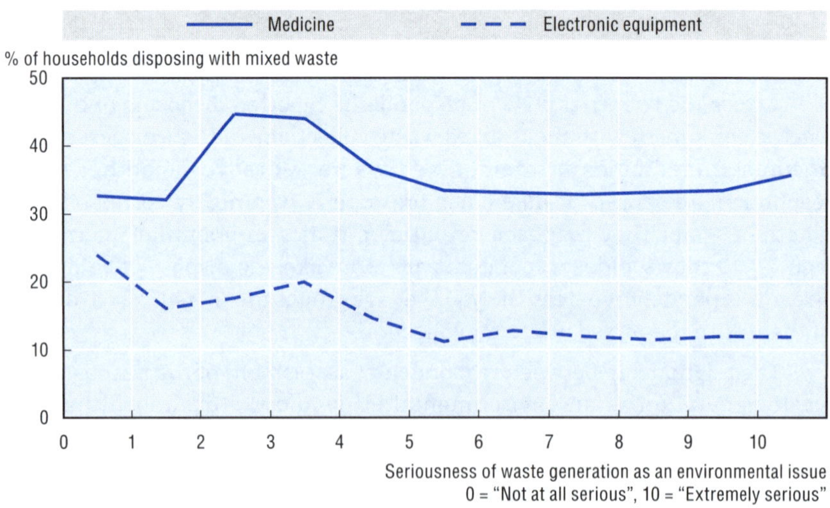

5. Waste separation and recycling

Respondents were also asked to indicate the percentage of the following five materials they separated: food and garden waste, glass bottles and containers, metal cans (aluminium, tin and steel), paper and cardboard, plastic bottles and containers.

For each of these categories of waste, respondents were asked if the following collection services were available:

- Bring back recyclable materials (to store/manufacturer) without a refund
- Bring back recyclable materials (to store/manufacturer) with a refund
- Drop-off centre available for recyclable materials
- Door-to-door (curb side) service available for picking up recyclable waste.

Figure 7.12 shows the breakdown of how many of the different materials are separated by the respondents. The overall average was also calculated in order to compare with the results from the 2008 EPIC Survey (OECD, 2011).

Figure 7.12. **Average number of different materials separated**

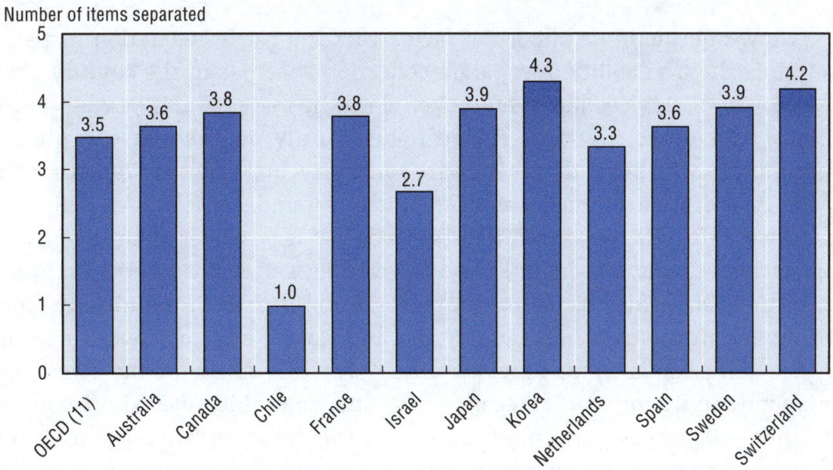

There is an increase in the number of different materials separated compared to the previous study, with an average of 2.5 materials separated per household in 2008 and 3.5 materials separated per household in the current 2011 data. Looking at the separation rates by age, it appears that 18 to 24 year-olds are least likely to separate their waste with 19% choosing not to do so; 16% of 25 to 34 year-olds choose not to separate, whereas the figure is 14% for the 35-44 age group and 13% for older respondents. In addition, the results suggest that respondents most concerned by environmental issues are much less likely to report that they do not recycle the different materials at all (Figure 7.13).

Figure 7.13. **Waste separation and importance of environmental concerns relative to other global issues**

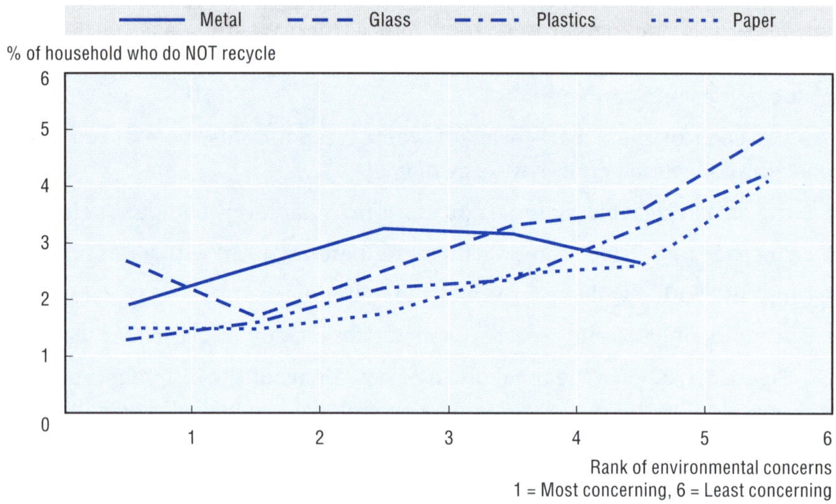

However, the most important factor affecting waste separation levels is certainly the availability of separate collection schemes in the country. The availability of services has been broken down by material and by country in Figure 7.14. Note that this figure focuses only on the sub-sample of respondents who knew that there were some recycling services available for a given type of waste, which was over 90% of respondents for most countries and most categories of waste. There is a wide variation across all the countries in which materials are separated and recycled, as well as the schemes in place to offer this facility. Door-to-door programmes are the dominant type of service in Australia, Canada, Japan and Korea, and to a lesser extent in the Netherlands. In the other countries, drop-off centres are the dominant recycling service for most categories of waste. Availability of drop-off centres for metal wastes is particularly strong in Sweden and Switzerland. Chile appears to have the least availability of recycling/separation services.[4]

With respect to deposit-refund systems (Figure 7.15), there is a vast difference from one country to another between what is available. Israel uses this system extensively to encourage recycling of glass and, to a lesser extent, metal. Canada, Korea, the Netherlands and Sweden also use deposit-refund systems, unlike countries such as France or Spain.

Figure 7.14. **Households using recycling services, by country and by service type**

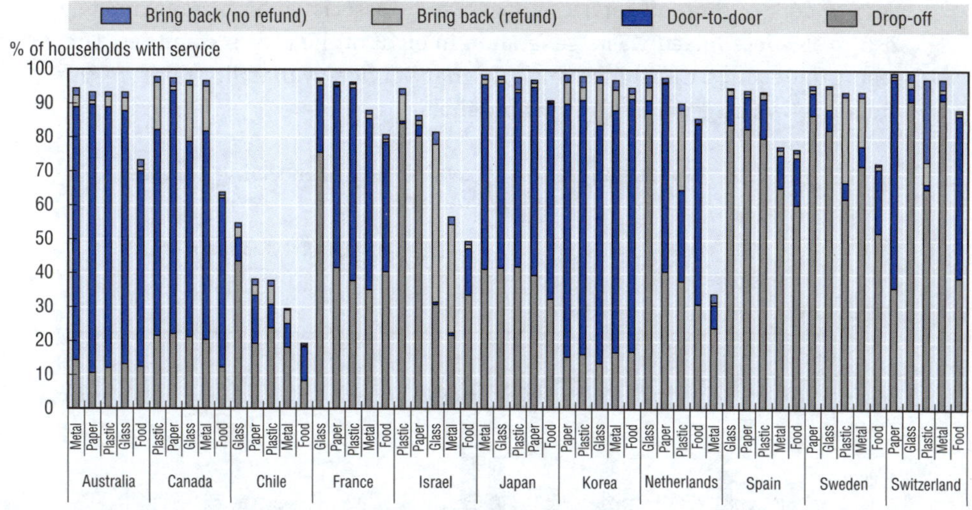

Note: Calculated for respondents who stated knowing that recycling services are available to them. Respondents who did not know whether such services are available are excluded from this figure, but are shown in Figure 7.19.

Figure 7.15. **Households who stated having deposit-refund systems**

Among those respondents knowing the service is available

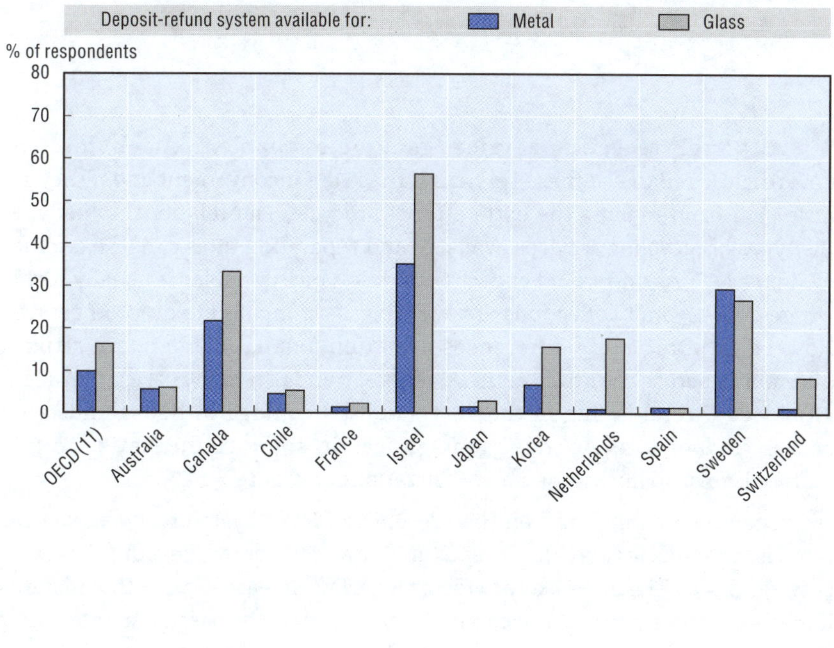

The availability of a separation service has a significant impact on the rate at which respondents separate waste, as Figure 7.16 shows. The availability of door-to-door or drop-off services appears to be associated with approximately 28% to 45% less mixed waste generation in eight of the eleven countries. This result underlines the significant role that service availability plays in the amount of waste generated by the household.

Figure 7.16. **Waste generation by availability of recycling services and by country**

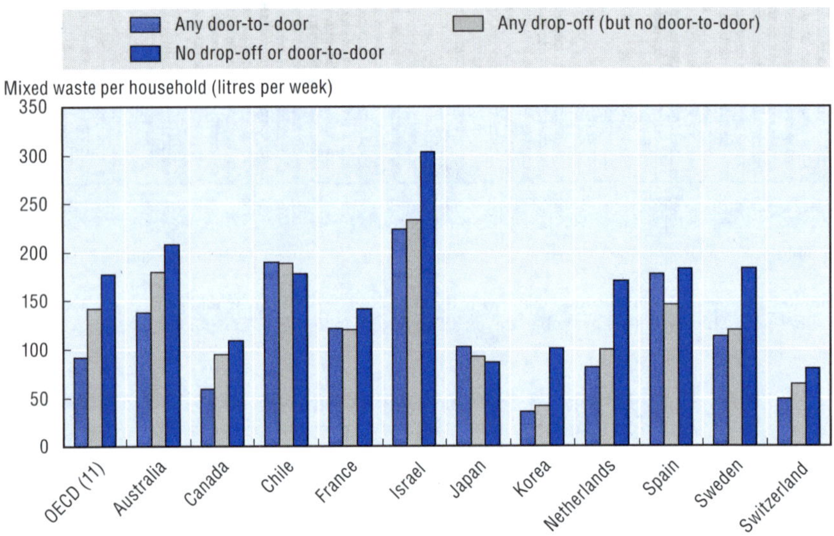

Although recycling services may be available, households may nevertheless fail to use these services if these are inconvenient and if they see little gain from making the extra effort to recycle. The relationship between the types of recycling services available and separation rates can be examined in Figure 7.17. As one would expect, separation of recyclables is usually higher where door-to-door collections are available, as compared to drop-off centres. However, the recycling differences are often small, but there are striking differences across countries in these general trends. In many cases, the lack of an obvious trend is due to the very small percentage of households with access to door-to-door collection services in some countries for certain materials (e.g. Spain, Sweden and Switzerland in Figure 7.14).

The relationship between the presence of PAYT systems and separation rates is, as expected, small: for each material considered, respondents were between 3% and 4% more likely to report recycling at least some of that material if they were subject to a PAYT scheme (volume- or weight-based). An interesting avenue for further research is the analysis of the relative importance of PAYT

Figure 7.17. **Household separation rates by type of waste, service availability, and by country**

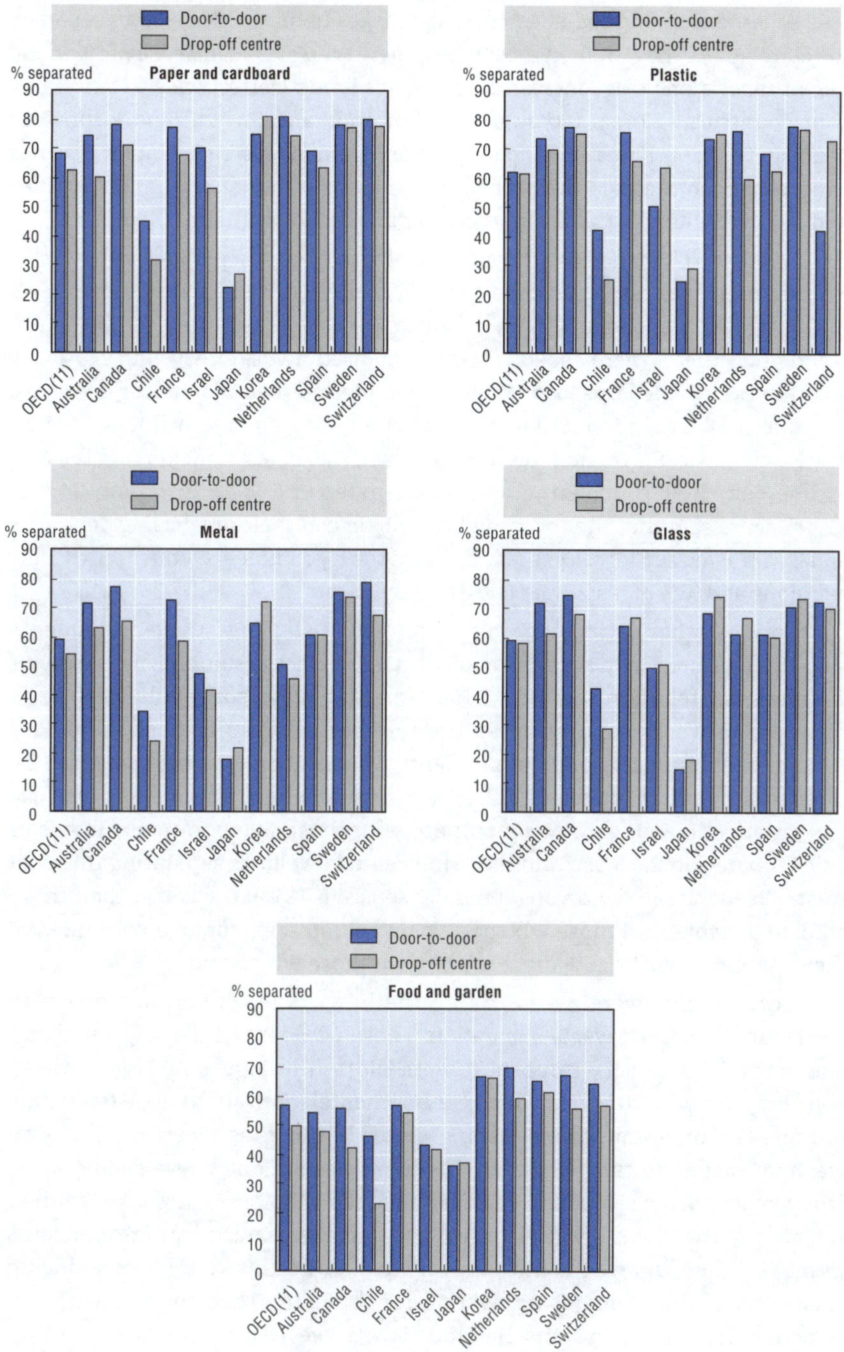

schemes and collection services on recycling rates, and more particularly the conditions under which they are complementary or substitute policies.

When examining the effect that age makes upon the decision to separate, the correlations show that, in most cases, there is an overall increase of 5% in the rate of separation as age increases by one age group (corresponds to ten years). Four countries – Korea, Japan, the Netherlands and Switzerland – show an opposite trend: in Korea and Japan, food and garden waste separation declines with respondents' age. In the Netherlands, paper and plastic separation declines with age. In Switzerland, glass separation rates decline with age.

Another tool to increase the amount of waste being separated and recycled is to use more waste-specific logos on products in places where they will be seen. In this survey, respondents in only four of the eleven countries – Chile, Israel, Japan and Korea – were asked if they recognised a waste recycling eco-label, whether they trusted this logo or whether it affected their purchasing decisions. In Chile, 77% of respondents said that the recycling logo influenced their purchasing, and 88% trusted this logo. In Israel, 70% of respondents said the logo influenced their purchasing and 91% trusted the logo. In Japan, 34% of respondents said that the recycling logo influenced their purchasing decisions, and 87% trusted the logo. In Korea, 56% said the logo influenced their purchasing decisions and 93% of respondents trusted the logo.

Various other socio-economic factors were also investigated. Gender is slightly correlated with separation of wastes: across the sample, males recycle 1% more wastes than females. Education also plays some role, with a 2% decrease in waste separation as the number of years of post-secondary education increases by one. Respondents who are home-owners separate 8% more waste than those tenants. The type of area where respondents live also has an effect on the rate of separation, with those living in major towns or cities separating the least, and those in isolated dwellings separating the most waste. Respondents who vote generally separate 18% more waste than those who do not vote, and those who partake in volunteering for an environmental organisation separate 15% more waste than those who do not.

Respondents were asked to rate the importance of various factors in motivating increased waste separation. Those who replied that a given factor was important to very important were plotted in Figure 7.18. Responses revealed a general pattern in which environmental motivations were frequently indicated as important, whereas being viewed by others as a responsible citizen was the least important. However, this trend was strikingly reversed in Israel, where respondents seemed to consider all of the factors less important in their recycling motivations. It was also observed that a large majority of respondents in most of the surveyed countries indicated "collecting refunds or reducing waste charges" as a major motivation for recycling. This, coupled with the evident lack of PAYT in all but a handful of the surveyed countries, suggests that

Figure 7.18. **Factors motivating households to separate waste**

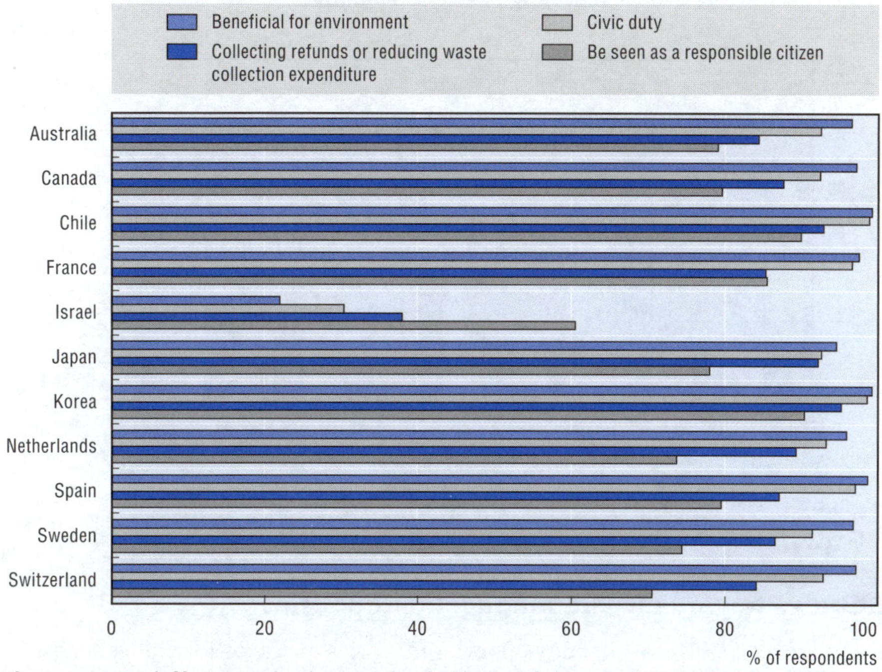

Note: The responses varied between 0 (not important) and 10 (extremely important). In this figure, only the percentage of responses that were above 5 (the midpoint) for each of the categories is presented.

the possibility of collecting refunds from recycling efforts was an important motivating factor in households' recycling behaviour.

These results show that the main factor that motivates separation is the issue of being beneficial to the environment, whereas being seen as a responsible citizen is generally not ranked as an important a factor. Among the countries surveyed, the role of civic duty is ranked the highest in Chile and Korea, followed by Spain and France. Once more, further empirical analysis would be beneficial to further develop the relationship between motivation and waste separation.

Although the nature of the motivation to recycle has a significant impact on whether a household separates or not, it is also dependent upon awareness of the services available. Self-assessed levels of awareness about recycling options are shown in Figure 7.19.

There are several policy relevant factors that affect households' intention to separate their wastes, such as general environmental concern and attitudes, and further in-depth analysis will be required to understand their relationship more fully. Further empirical analysis should be undertaken to examine the causality (the *effect* of service availability on waste generation), incorporating other demographic, economic and policy factors.

Figure 7.19. **Households who state not being informed about recycling services availability**

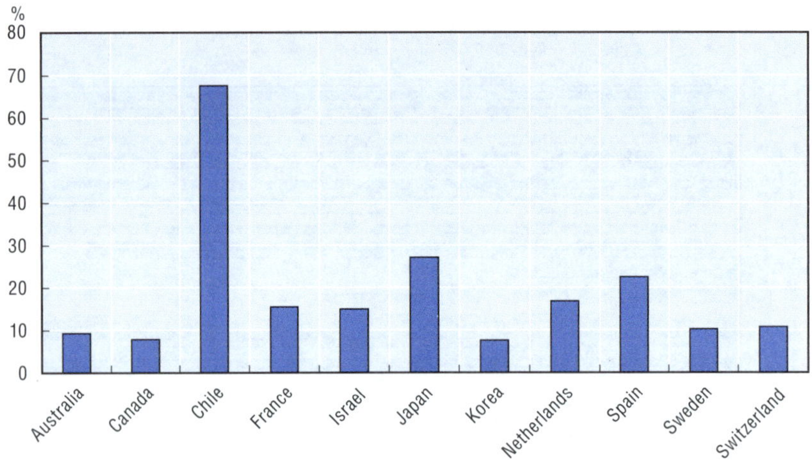

Note: Responses varied between 0 (not at all informed) and 10 (very well informed). In this figure, those indicating 0 to 3 are classified as not being informed.

6. Attitudes towards waste management policies

Increasing the availability of drop-off centres and door-to-door collection of recyclables gives households a convenient means to reduce the waste they send to landfills, and – depending on the billing system in place – to save money. As seen above, both recycling programme availability and the use of unit-based charges are strongly correlated with reduced mixed waste generation.

However, to better evaluate the potential of these policies, it is important to know households' level of support for them. To investigate this, respondents were asked to what extent they would support various government actions to reduce household waste generation. Responses ranged from 0 (don't support) to 10 (strongly support). Figure 7.20 shows the fraction of respondents who indicate a high level of support – greater than or equal to 7 – on the 10-point scale.

Respondents appear to be most supportive of measures which incite manufacturers to reduce the packaging they supply with their products, and measures that encourage people to buy products with less packaging. Educating the public is also seen as an important factor, particularly in Chile, Israel and Spain. Less than 46% of respondents support charging by weight or volume for waste collection, even though these policies were shown above to be effective at reducing waste generation. Given the direct and transparent financial implications for households of unit-based fees, such a

Figure 7.20. **Households' support for five different waste-reduction policies**

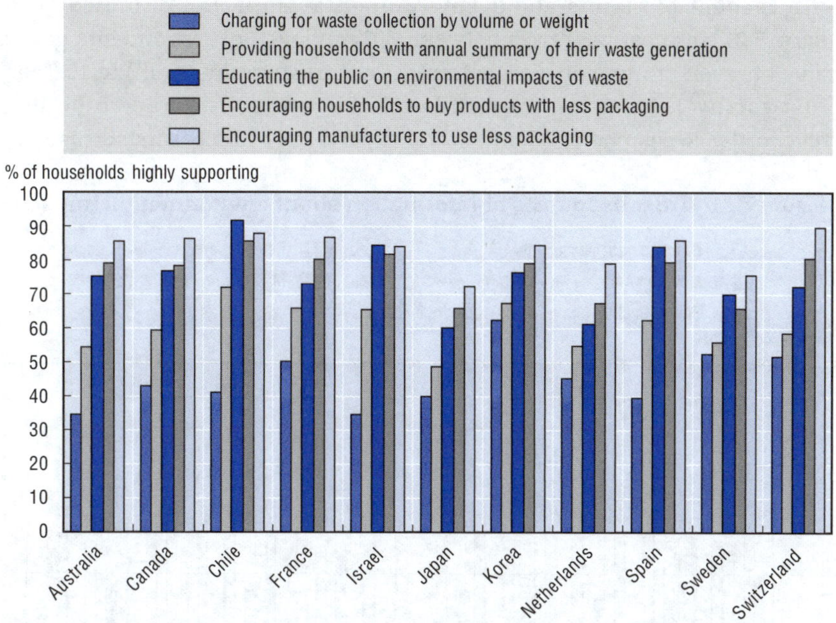

response is not surprising. However, it must be remembered that the average costs of the other options may be equally or more costly for households, if less evident.

Looking at how support for weight- or volume-based waste charges vary by respondent characteristics, one finds that 35 to 44 year-olds were most in support of a PAYT system, and that the 18 to 24 year-old group were least in support of such a policy. There is no correlation with gender, both males and females being equally in support of waste charging. Respondents living in isolated areas were most in support of PAYT compared to those in suburban areas who were the least supportive (5.6 compared to 5.3 out of 10). Curiously, respondents who are in general happier with their life (as measured by a question on perceived well-being) are also more in support of PAYT.

Respondents' relative concern for environmental issues is also positively associated with support for PAYT: those respondents who were more concerned with the environment are more likely to support the government imposing weight- or volume-based tariffs on waste collection services. In a similar scenario, the same was found when assessing respondents' opinions on the seriousness of waste generation and their general concern for the environment. Voters are more likely to support waste charging as are those who volunteer for an environmental organisation. These hypothesised relationships merit further empirical analysis.

Ensuring support for policies depends upon respondents' degree of trust with respect to claims about environmental impacts. As indicated in Figure 7.21, in most countries between 30% and 40% of respondents gave a value of seven or more on a scale from 0 ("not trustworthy at all") to 10 ("very trustworthy"). Manufactures and retailers are considered to be less trustworthy. Researchers, scientists and experts are by far the most trusted.

Figure 7.21. **Trust in sources of information about environmental impacts**

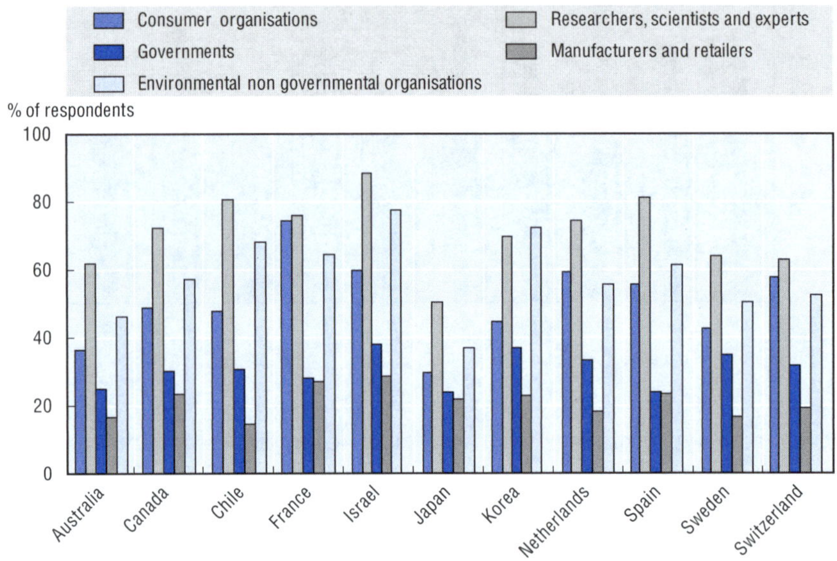

7. Conclusions

This analysis confirms that there are numerous factors determining households' waste prevention and recycling decisions, including preferences for consumption and environmental quality, socio-economic constraints (e.g. income and household location), as well as the policy setting (how tariffs for waste collection are levied). For example, the more people in the household, the more waste the household will generate, as expected, but the increase is less than proportional. Moreover, country variations are significant.

Waste generation also appears to vary with respondents' age, with an average of 38% difference between the volume of waste generated by the age group producing the least and the age group producing the most. The age group producing the most waste varies by country, but appears to be mainly the 18 to 24 age group or the 55+ age group. More years of post-secondary education is tied to a decrease in the amount of waste produced. This would suggest that new waste-related policies need to be targeted on specific household groups to improve the effectiveness of the measures. Household income is an important factor in the generation of waste, and this is an area requiring further empirical analysis: the relationship between household disposable income, on the one hand, and both waste generation and the willingness to separate waste, on the other hand, is a subtle but important issue for effective waste management policies.

The prevalence of PAYT billing systems remains low in all but a handful of the countries surveyed. This illustrates a policy area with high potential, since it was shown above that waste generation tends to be between 20% and 30% lower with such systems, and that recycling is somewhat higher.

However, household separation of waste depends most strongly on the presence and quality of recycling services. Comparing door-to-door collection and drop-off centres, the more convenient the service, the more likely the household is to recycle. This gain in recycling rates with such services, particularly with regard to door-to-door services, should be compared to the costs of implementing such systems in order to evaluate their cost-effectiveness at diverting waste from being sent to landfills. Respondents' characteristics – in particular their age – were also found to be associated with recycling rates, with an increase in recycling as age increases. Again, the reasons for this descriptive pattern deserve further empirical analysis.

The role of environmental concerns in waste prevention and recycling decisions also deserves more in-depth empirical analysis. A positive association was found between the overall level of concern for the environment and reduced waste generation. Furthermore, specific concerns about waste generation relative to other environmental issues were found to be positively associated with separation rates in countries where recycling systems are available. Because households' attitudes both influence and are influenced by a number of factors, the results described here deserve further empirical analysis to assess whether there is a causal link between attitudes and behaviour.

Separation of hazardous waste is an issue which stands out as requiring further government intervention, especially with respect to old and unused medicines – which is an area the survey has fortunately collected data on. There appears to be a lack of awareness as to how to dispose of these items correctly, with an overall average of 34% of households disposing of medicines with their mixed wastes. This would indicate a need for further information-based measures within new waste related policies.

Awareness of the services availability plays an important part in household separation and recycling, with an overall average of 19% of households not knowing what is recyclable in their area. If households are informed about what is available and about the environmental and economic benefits of separating and recycling mixed wastes, there is likely to be an increase in separation of wastes and a decrease in the generation of mixed wastes. This suggests a role for policies providing education to households about available recycling services, a policy which indeed was one supported by a majority of households in all countries.

There is a relatively low level of trust in information provided by governments about the environmental impacts of products. Researchers, scientists and experts could perhaps be consulted more effectively in public awareness campaigns, as results indicate that these are the most trusted rather than manufacturers, retailers and governments. This was a key area that respondents thought to be a way forward with respect to separation and recycling. Households also expressed a wish for manufacturers to take some responsibility on waste prevention, by reducing the amount of packaging on their products and using recyclable packaging where necessary. Future rounds of the survey may examine whether respondents would accept some price premium on products that utilise more recycled or recyclable packaging.

Prominent eco-labelling on packaging may also help households understand how it can be recycled and how much packaging is used in the sale of products.[5] There was strong support for increased provision of such information. However, when it comes to the effectiveness of recycling information on labels, it is important to note that respondents who currently separate and recycle more than average are already most interested in this option, another area for further analysis.

The results of this survey indicate that current waste policies are becoming successful in diverting a proportion of valuable materials from landfills and reducing environmental impacts. These policies need to go further by using a targeted approach and educating the public while offering incentives to manufacturers to use less packaging. More efficient collection schemes need to be implemented either by national or regional governments, local municipalities or by private companies to ensure effective collection of separated materials. In terms of effectiveness in reducing waste generation, there is also clearly a role for unit-charging systems (i.e. PAYT). Improved understanding of their positive environmental and financial (public and private) implications is essential.

Notes

1. Phenomenon encountered in Seattle when the city imposed an important landfill tax increase and sometimes referred to as the "Seattle Stomp" since then.
2. For example respondents were asked to indicate whether they strongly disagree (-2), disagree (-1), no opinion (0), agree (1) or strongly agree (2) with the following statement: "I am willing to make compromises in my current lifestyle for the benefit of the environment" (Q26). The responses were normalised to give a scale from -2 to +2, with higher values indicating more "pro-environmental" attitudes.
3. Old batteries and accumulators are other examples.
4. It is important to note that for some respondents (e.g. in apartment buildings) there may be a degree of confusion about the distinction between on-site and drop-off collection services.
5. Clearly indicating that some products need to be collected separately is also important (e.g. domestic batteries).

References

Åberg, H. et al. (2009), "Inconsistent pathways of household waste", *Waste Management*, Vol. 29, pp. 1798-1806.

Adams, R. M., S. Hong and H.A. Love (1993), "An Economic Analysis of Household Recycling of Solid Wastes: The Case of Portland, Oregon", *Journal of Environmental Economics and Management*, Vol. 25, pp. 136-146.

Allers, A. M. and C. Hoeben (2010), "Effects of Unit-Based Garbage Pricing: A Differences-in-Differences Approach", *Environmental and Resource Economics*, Vol. 45, pp. 405-428.

Afroz, R., K. Hanaki and R. Tuddin (2010), "The Role of Socio-Economic Factors on Household Waste Generation: A Study in a Waste Management Program in Dhaka City, Bangladesh", *Research Journal of Applied Sciences*, Vol. 5, pp. 183-190.

Ashenmiller, B. (2011), "The Effect of Bottle Laws on Income: New Empirical Results", *The American Economic Review*, Vol. 101, pp. 60-64.

Badruddin M. Y., F. Othman, N. Hashim and N. C. Ali (2002), "The role of socio economic and cultural factors in municipal solid waste generation. A case study in Taman Perling Johor Bahru", *Journal of Technology*,Vol. 37, pp. 55-64.

Bandara, N. et al. (2007), "Relation of waste generation and composition to socio-economic factors: a case study", *Environmental Monitoring and Assessment*, Vol. 135, pp. 31-39.

Berglund, C., O. Hage and P. Söderholm (2009), "Norms and economic motivation in household recycling: Empirical evidence from Sweden", *Resources, Conservation and Recycling*, Vol. 53, pp. 155-165.

Best, H. (2009), "Structural and Ideological Determinants of Household Waste Recycling: Results from an Empirical Study in Cologne, Germany", *Nature and Culture*,Vol. 4, pp. 167-190.

Blume, D. et al. (1994), "Market-based incentives and residential municipal solid waste", *Journal of Policy Analysis and Management*, Vol. 13, pp. 681-698.

Bruvoll, A. and K. Nyborg (2004), "The Cold Shiver of Not Giving Enough: On the Social Cost of Recycling Campaigns", *Land Economics*,Vol. 80, pp. 539-549.

Bucciol, A., N. Montinari and M. Piovesan (2011), Do Not Trash the Incentive! Monetary Incentives and Waste Sorting, *Harvard Business School Research Paper Series*, Harvard.

Callan, S. J. and J.M. Thomas (1997), "The Impact of State and Local Policies on the Recycling Effort", *Eastern Economic Journal*, Vol. 23, pp. 411-423.

Callan, S. J. and J.M. Thomas (2006), "Analyzing Demand for Disposal and Recycling Services: A Systems Approach", *Eastern Economic Journal*, Vol. 32, pp. 221-240.

Dahlén, L. and A. Lagerkvist (2010), "Evaluation of recycling programmes in household waste collection systems", *Waste Management & Research*, Vol. 28, pp. 577-586.

Dahlén, L. et al. (2007), "Comparison of different collection systems for sorted household waste in Sweden", *Waste Management*, Vol. 27, pp. 1298-1305.

Duggal, V., C. Saltzman and M. Williams (1991), "Recycling: An Empirical Analysis", *Eastern Economic Journal*, Vol. 17, pp. 351-358.

Ebreo, A. and J. Vining (1990), "What Makes a Recycler?", *Environment and Behavior*, Vol. 22, pp. 55-73.

Ferrara, I. and P. Missios (2005), "Recycling and Waste Diversion Effectiveness: Evidence from Canada", *Environmental and Resource Economics*, Vol. 30, pp. 221-238.

Fujiwara, T., N.P. Thanh and Y. Matsui (2010), "Household solid waste generation and characteristic in a Mekong Delta city, Vietnam", *Journal of Environmental Management*, Vol. 91, pp. 2307-2321.

Fullerton, D. and T.C. Kinnaman (1994), "Household Responses for Pricing Garbage by the Bag", *NBER Working Paper Series*, No. 4670.

Fullerton, D. and T.C. Kinnaman (2000), "Garbage and Recycling with Endogenous Local Policy", *Journal of Urban Economics*, Vol. 48, pp. 419-442.

Hong, S. (1999), "The effects of unit pricing system upon household solid waste management: The Korean experience", *Journal of Environmental Management*, Vol. 57, pp. 1-10.

Jenkins, R. R. (1993), *The economics of solid waste reduction: The impact of user fees*, E. Elgar,Vermont, United States.

Jenkins, R. R. et al.(2003), "The determinants of household recycling: a material-specific analysis of recycling program features and unit pricing", *Journal of Environmental Economics and Management*, Vol. 45, pp. 294-318.

Joshi, S.V., S.F. Sidique and F. Lupi (2010), "Factors influencing the rate of recycling: An analysis of Minnesota counties", *Resources, Conservation and Recycling*,Vol. 54, pp. 242-249.

Karbassi, A. et al. (2012), "The effects of socioeconomic parameters on household solid-waste generation and composition in developing countries (a case study: Ahvaz, Iran)", *Environmental Monitoring and Assessment*, Vol. 184, pp. 1841-1846.

Mazzanti, M. and R. Zoboli (2009), "Municipal Waste Kuznets Curves: Evidence on Socio-Economic Drivers and Policy Effectiveness from the EU", *Environmental and Resource Economics*, Vol. 44, pp. 203-230.

Meneses, G. D. and A.B. Palacio (2005), "Recycling Behavior", *Environment and Behavior*,Vol. 37, pp. 837-860.

Nixon, H. et al. (2006), "Household Willingness to Recycle Electronic Waste", *Environment and Behavior*, Vol. 38, pp. 183-208.

OECD (2008), OECD Environmental Outlook to 2030, OECD Publishing. doi: *10.1787/9789264040519-en.*

OECD (2011), *Greening Household Behaviour: The Role of Public Policy*, OECD Publishing. doi: *10.1787/9789264096875-en.*

Reschovsky, J. D. and S.E. Stone (1994), "Market incentives to encourage household waste recycling: Paying for what you throw away", *Journal of Policy Analysis and Management*, Vol. 13, pp. 120-139.

Tam, V. W. Y. and C.M. Tam (2007), "Waste reduction through incentives: a case study", *Building Research and Information*, Vol. 36, pp. 37-43.

Greening Household Behaviour
Overview from the 2011 Survey

Chapter 8

Household attitudes across environmental domains and time

by
Nick Johnstone, Zachary Brown *and* Ysé Serret-Itzicsohn*

This concluding chapter analyses selected issues cutting across the thematic areas examined in the survey (energy, transport, water, food and waste) and time. It looks at how willingness-to-pay patterns differ from one environmental good to the other (e.g. electric cars, "green" electricity and organic food). It also examines motivations to conserve energy and water. In addition, the chapter provides a comparison of some questions for the six countries involved in the two rounds of the Survey on Environmental Policy and Individual Behaviour Change (EPIC).

* OECD Secretariat, Environment Directorate.

The statistical data for Israel are supplied by and under the responsibility of the relevant Israeli authorities. The use of such data by the OECD is without prejudice to the status of the Golan Heights, East Jerusalem and Israeli settlements in the West Bank under the terms of international law.

The previous five thematic chapters have focused on households' attitudes and choices in single environmental domains: energy, water, transport, food and waste. This concluding chapter looks at selected issues cutting across areas. These include questions relating to willingness-to-pay (WTP) for environmental goods and services, motivations to conserve resources, label recognition and the adoption of new technologies. Emerging patterns are examined, as well as the role of norms and attitudes for the design of policies.

Also presented here is evidence on differences in responses to selected questions among the six countries that participated in both rounds of the survey: Australia, Canada, France, Korea, the Netherlands and Sweden. However, when comparing the 2008 and 2011 surveys, it is important to bear in mind that the samples are not the same, even if the sampling procedure was. Nonetheless, the review provides interesting insights into changing attitudes and concerns.

1. Willingness-to-pay for different "environmental" goods

There were three questions on willingness-to-pay (WTP) pertaining to the energy, food and transport domains. While the respective thematic chapters analyse data for each of these WTPs in isolation, it is interesting to examine patterns of responses across the different environmental goods examined: Three similarly structured questions were asked about WTP for organically grown fruit and vegetables (see Question 88 in Annex A), the provision of green" energy (Q71) and electric cars (Q52). The response format was a slider bar, running from 0% to 100% for the energy and car questions. In the case of food, respondents could indicate as much as 200% since, according to pre-tests, a small minority of respondents answered that they would be willing to pay more. This distinction is important to keep in mind.

Rather than the absolute values, it is interesting to note the different patterns across the different goods in different countries (Figure 8.1). For instance, in the Netherlands the median WTP for electric cars is 30% while it is 2% for renewable energy. In Switzerland, the two values are the same (10%). In four countries, WTP for organic food is greater than WTP for an electric car (France, Korea, Sweden and Switzerland). The previous thematic chapters provide detailed reasons for these differences, but clearly factors such as assumptions about the baseline characteristics of the conventional substitute in each country (such as the percentage of renewables already in the supply

Figure 8.1. **Median willingness-to-pay for different environmental goods**

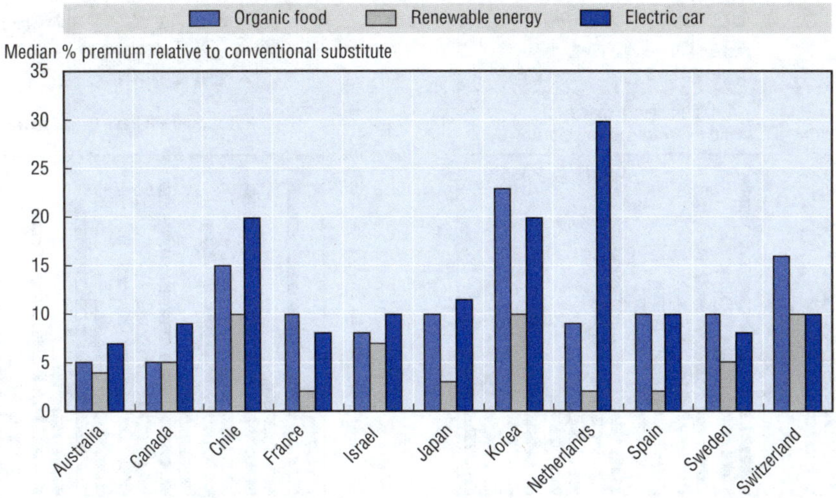

mix), geographical issues (population density for electric vehicles) play a role. However, differences in underlying preferences also play a prominent role.

When looking across domains, the reasons given by those respondents who were not willing to pay extra for these goods, those stating WTP=0, are quite revealing. The following four figures report on four different reasons given. Respondents were able to indicate multiple factors as reasons why they were not willing to pay more for the good in question. Figure 8.2A shows the percentage of respondents who indicated that if they had more money they would be willing to pay more for the three goods. This would indicate that income constraints were an important factor in their choice. While the numbers are lower than for other potential responses (see below), they are still quite high (30% to 50%) despite the fact that they could theoretically indicate a value as low as 1%. There is no great variation across goods and between countries.

The pattern is very different for those who responded that they felt they should not have to pay extra for the good in question (Figure 8.2B). In all countries, and across the three goods considered, it is for electric cars that this reason is more frequently provided as a reason for not being willing to pay extra. This may, of course, have something to do with the perception that electric cars are "inferior" private goods. However, it is important to note that respondents could also indicate that their WTP was zero because they preferred conventional internal combustion engine vehicles. At the other extreme, respondents were least likely to cite this as justification for not being willing to pay for organic fruit and vegetables.

Figure 8.2. **Reasons for not being willing to pay extra**

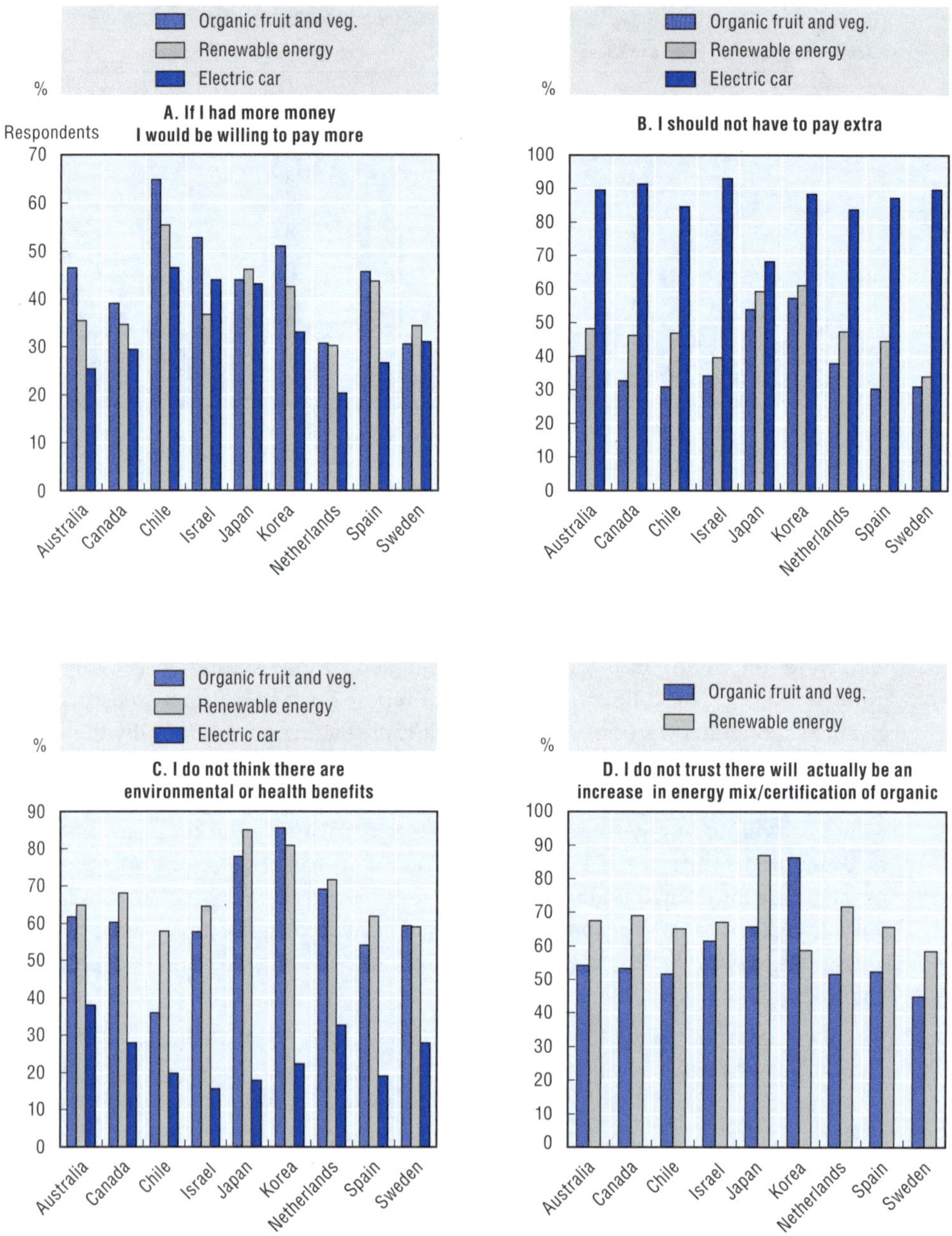

It can be argued that those who really have WTP = 0 (as opposed to those who specify WTP = 0 as a "protest" response) can be gauged from whether or not respondents believe the good does not generate environmental benefits (or health benefits in the case of organic food). In this case, the percentage of respondents is remarkably similar for organic fruit and vegetables and renewable energy, and much greater than for electric cars. Stated differently, perception of environmental benefits was not important in determining whether or not people would pay extra for an electric car.

And finally, even if some respondents believe that there may be potential environmental benefits associated with purchasing the good, they may not "trust" that these benefits will actually occur (Figure 8.2D). This type of response is most relevant for credence goods whose environmental benefits are expressed by experts or through certification systems. For example, respondents – while perhaps desiring more renewables in the energy mix – may not think that their money would be used effectively if they paid more for a "green" option on their utility bills. This type of reason for specifying a WTP=0 can be considered as a form of "protest" response, though it is likely that if they were given the actual (as opposed to hypothetical) choice, they would still choose not to purchase the good, since they do not trust its alleged benefits. Predictably, this reason was also frequently cited for not being willing to pay extra for organic food. Korea is the one case where this was a more important factor for organic fruit and vegetables than for renewable energy.

2. Reported motivations to conserve resources (energy and water)

Respondents indicated what factors were most likely to encourage them to conserve resources in two domains: energy and water (see Q79 and Q96). Importantly, the structure of the questions were identical, both in terms of wording* and the nature of the response (a scale from 0 = not important to 10 = very important). The pattern is very similar across the two domains (see Figure 8.3). Higher energy or water prices are reported to have the least impact, and less expensive equipment the greatest impact, clearly reflecting a strategic bias in the responses. That said, the proportion of respondents who recognised the important role of higher prices for energy and water is not negligible.

Setting these pecuniary factors aside, there are some differences between perceived effectiveness of factors to promote water and energy conservation. For instance, information on the environmental impacts of using resources is

* "How important would the following factors be in encouraging you to reduce your household's energy (Q79) or water (Q96) consumption?".

Figure 8.3. **Reported effect on resource use**

(Scale normalised to mean = 1.0)

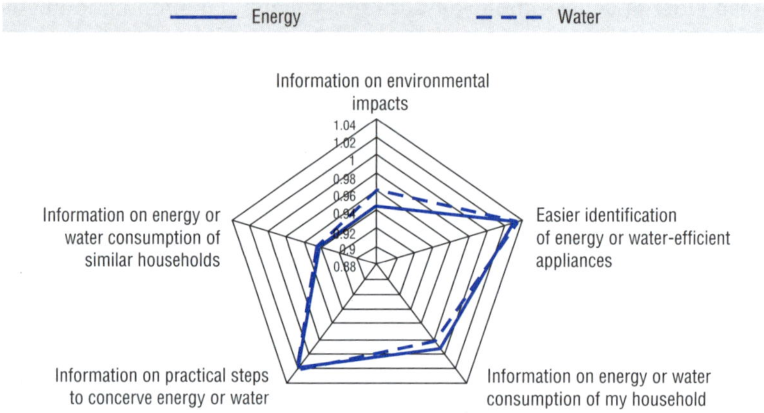

reported to have a greater effect on water use than on energy use. Conversely, better information on the households' consumption is thought to have a greater impact on energy consumption.

Figure 8.4 presents the number of respondents who reported different factors as being important in the two cases. As noted, respondents can provide responses from 0 (not important) to 10 (very important). In the case of higher resource prices, there is a large percentage at the two extremes. In addition, there are relatively large numbers of respondents who indicate that prices would have an effect on one resource but not on the other. Interestingly, this remains the case even if one focuses on the sub-sample of respondents who pay

Figure 8.4. **Factors that would encourage reduced water or energy consumption**

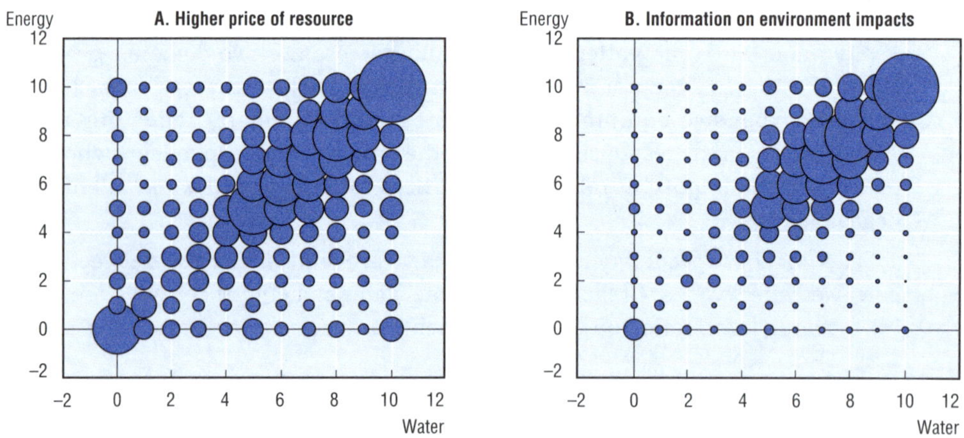

for both resources on a per-unit basis. As far as the provision of information on environmental impacts is concerned, there is a much higher percentage on the diagonal, indicating that respondents feel that this will have a similar impact on both resources. This is also true of other areas examined (e.g. information on own consumption and/or that of similar households).

3. Recognition of labels

Respondents were asked a series of questions about recognition, understanding and use of eco-labels corresponding to a variety of environmental domains, ranging from labelling for high-efficiency appliances to sustainably caught fish. In the first instance, they were presented with an image of a label on-screen and requested to indicate if they recognised it. For those who did, they were then asked if they understood, trusted and used the label. Since many labels are country-specific, most of the data are not strictly comparable. However, the Marine Stewardship Council (MSC) label and the Energy Star labels were presented to respondents from the same nine countries. This is interesting since one label relates to a perishable good purchased on a regular basis (seafood), and the other to the occasional purchase of consumer durables (electric appliances).

As can be seen in Figure 8.5, mean recognition is much higher for the Energy Star than for the MSC label. Switzerland and Sweden are the outliers, with relatively high rates of MSC label recognition (20% to 30%). Yet, even for the Energy Star label, it is only in Canada and Australia where recognition exceeds 50% of respondents. It also interesting to note that cross-country

Figure 8.5. **Recognition of different labels**

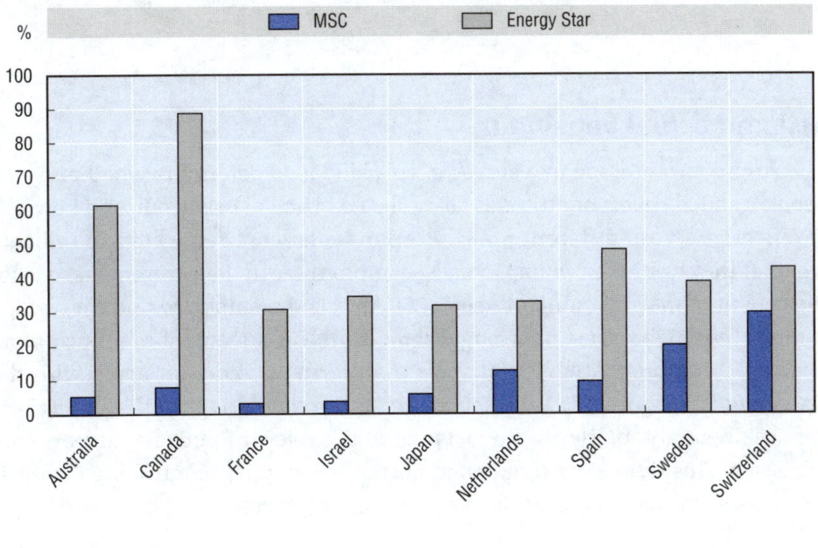

patterns move in sync between labels: countries with higher levels of Energy Star recognition tend to have higher levels of MSC recognition. This suggests that consumers may possess some underlying "label awareness" that is evidenced in the recognition of a variety of labels. Further empirical work is necessary to investigate this hypothesis.

Among those who recognise both MSC and Energy Star labels, the level of trust is greater for the MSC labels even though its recognition is so much lower (see Figure 8.6). This is true in all countries except Australia and France. Moreover, the level of reported use of the label is much higher for the MSC label than for the Energy Star label. The one exception is Canada.

Figure 8.6. **Trust and use of different labels**

Of those who recognise the label

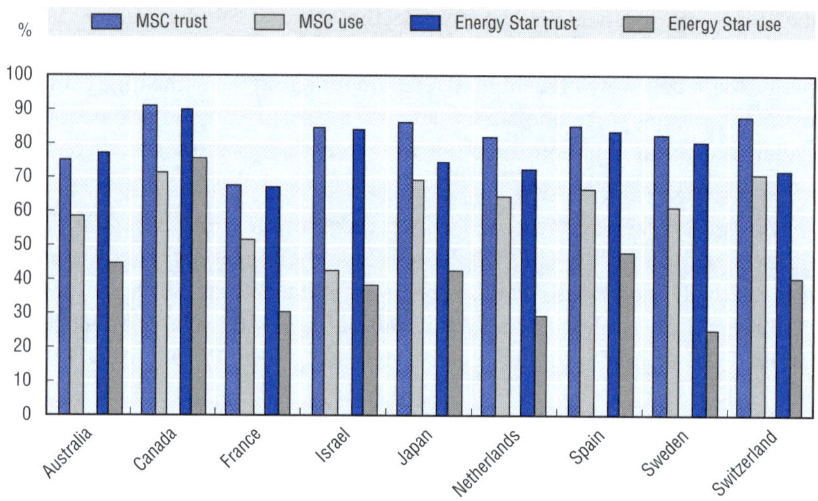

4. Stated and actual behaviour

An important issue in a survey of this kind is to determine whether or not people's stated beliefs or attitudes are reflected in actual behaviour, and how this is reflected across different areas. Respondents were requested to indicate whether they would be "willing to make compromises in their current lifestyle for the benefit of the environment" (see Q26). Among the employed and car-owning sample, those who agree or strongly agree with this statement are much more likely to commute by public transport than those who do not agree with the statement (see Figure 8.7). While other factors are clearly at play, the differences are striking, particularly as the distance to the relevant public transport stop increases. This difference remains important even when controlling for other factors (such as income, location of residence, and others).

Figure 8.7. **Commuting by public transport and willingness to compromise lifestyle**

Sample of employed and car-owning respondents

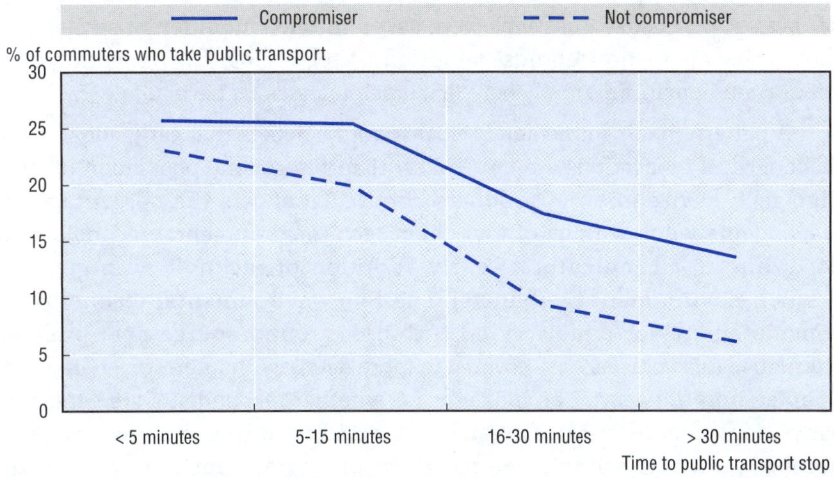

A similar relationship between attitudes and behaviour can be seen in terms of recycling. Among the sample of those who are not subject to a pay-as-you-throw (PAYT) scheme (to ensure that there are no financial benefits from recycling) and who do not have door-to-door collection of recyclables (thus requiring some effort on their part), those who agree with the statement have much higher reported recycling rates than those who do not (Figure 8.8).

Figure 8.8. **Recycling rates and willingness to compromise lifestyle**

Sample of households not subject to PAYT and no door-to-door collection of recyclables

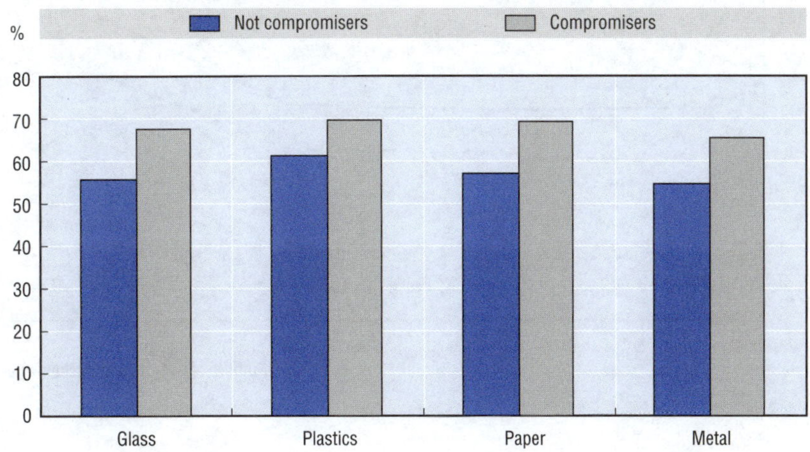

5. Households' adoption of technological innovations

Adoption of new technology can both increase and decrease households' environmental impacts. The nature of the effect depends on the technology adopted. A number of questions were asked throughout different sections of the survey about households' adoption of technological innovations, in general, and environmentally beneficial technologies, in particular.

A pattern that is immediately clear from the data is that early-adopters of technological innovations are more likely than the general population to own high-tech environmental products. Figure 8.9 shows the percentage of respondents who own one of four "high-tech" environmental technologies, according to an indicator of early adoption of technology in general (i.e. non-environmental products). The four environmental technologies considered – solar panels, wind turbines, ground-source heat pumps, alternative fuel vehicles – are covered in more detail in Chapter 3 on energy and Chapter 4 on transport. The indicator for whether respondents are generally early-adopters of technology at large was constructed by simply counting the number of the following three recent technologies which are present in households: flat screen televisions, mobile phones with internet connection, and internet telephony (e.g. Skype). Those who own all three innovative technologies are significantly more likely to own high-tech environmental products in particular.

Figure 8.9. **Respondents with high-tech environmental products by innovative technologies**

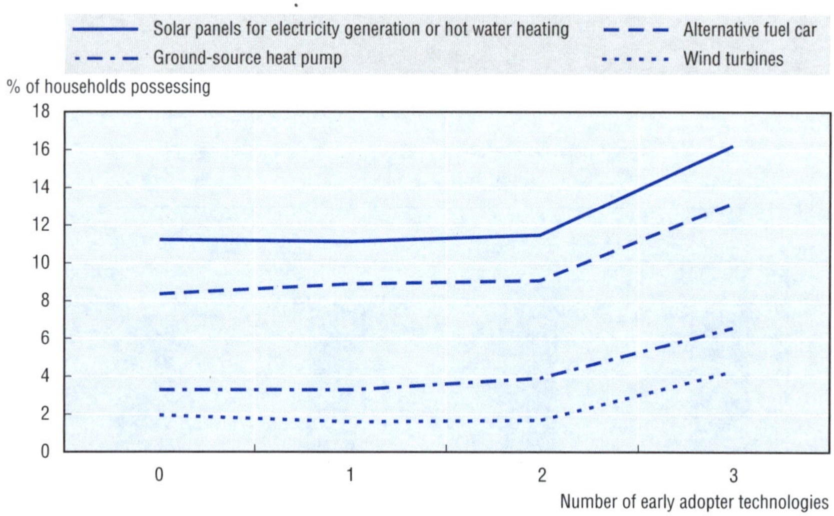

6. Comparison of selected responses from the 2008 and 2011 surveys

As noted, a number of questions were repeated in both the 2008 and 2011 surveys. To conclude this report, it is therefore interesting to compare changes in response patterns observed in the six countries common to the two surveys (Australia, Canada, France, Korea, the Netherlands and Sweden). The focus will be on some of the more general questions associated with environmental concerns and attitudes.

In both surveys, respondents were asked to rank "six issues facing the world today" in terms of their importance: international tensions, economic concerns, environmental concerns, health concerns, social issues and personal safety (Q22). The ranks have been inverted for ease of interpretation so as to allow comparison with other figures based on Likert and other scales – i.e. the further from the centre the greater the importance attached to the issue. As Figure 8.10 shows, the mean value for economic concerns is higher in 2011 than in 2008, indicating that respondents found it to be relatively more important. The opposite is true of environmental concerns. The biggest change is for personal safety (more important in 2008) and international tensions (more important in 2011).

Figure 8.10. **Average (inverted) rank of issues in 2008 and 2011 surveys**
(higher number = more concern)

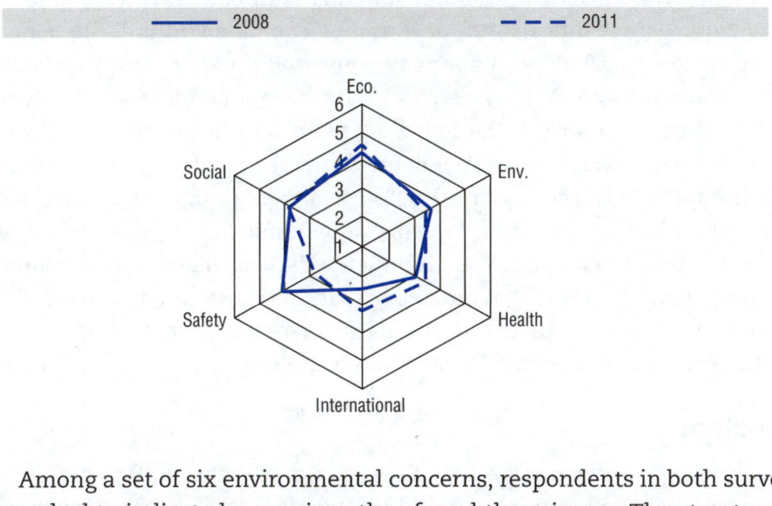

Among a set of six environmental concerns, respondents in both surveys were asked to indicate how serious they found these issues. The structure of the possible responses was different in the two cases; they have thus been normalised with a mean value of 1.0 and larger numbers indicating greater importance. The biggest changes are with respect to biodiversity and natural resource depletion, with the latter considered relatively more important among respondents in the 2011 survey, and biodiversity relatively more important in the 2008 survey (see Figure 8.11). Arguably, this is consistent with the inversion of ranks between economic and environmental concerns in Figure 8.10 since natural resource depletion has much more evident (although not necessarily important) economic implications.

Figure 8.11. **Mean reported seriousness of environmental concerns in 2008 and 2011 surveys**

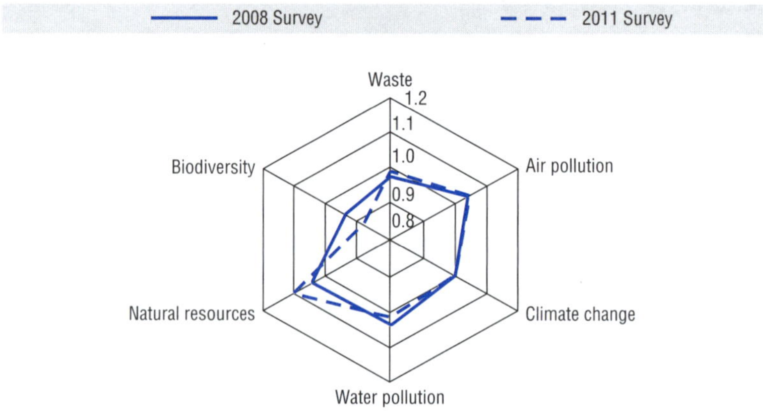

Respondents were also asked to state their degree of disagreement or agreement with specific attitudinal statements in both surveys. Three such statements were identical, and these can be compared. In order to generate Figures 8.12a to c below, responses were given values from -2 (strong disagreement) to +2 (strong agreement), with "no opinion" accorded a value of 0. Results are presented in the following figures. In five of the six countries (Korea being the exception), respondents expressed greater agreement in 2011 than in2008 with the statement that "environmental impacts are frequently overstated". In all countries except the Netherlands, they were less likely to agree with the statement that environmental issues will be resolved through technological progress. Most strikingly, in all six countries there was a significant increase in the percentage of respondents who felt that environmental issues should be dealt with primarily by future generations.

7. Conclusions

Comparing willingness-to-pay shows patterns that are different from one environmental good to the other and in underlying references. Respondents are generally willing to pay a higher premium for electric cars than for the provision of renewable energy. Surprisingly, the reason most often stated by respondents for not wanting to pay anything more for electric cars is that they considered that they should not have to pay extra, while income constraints are most often cited as determinant for organic food. The perception that

Figure 8.12. **Responses to selected attitudinal statements**

(2 = strongly agree, -2 = strongly disagree)

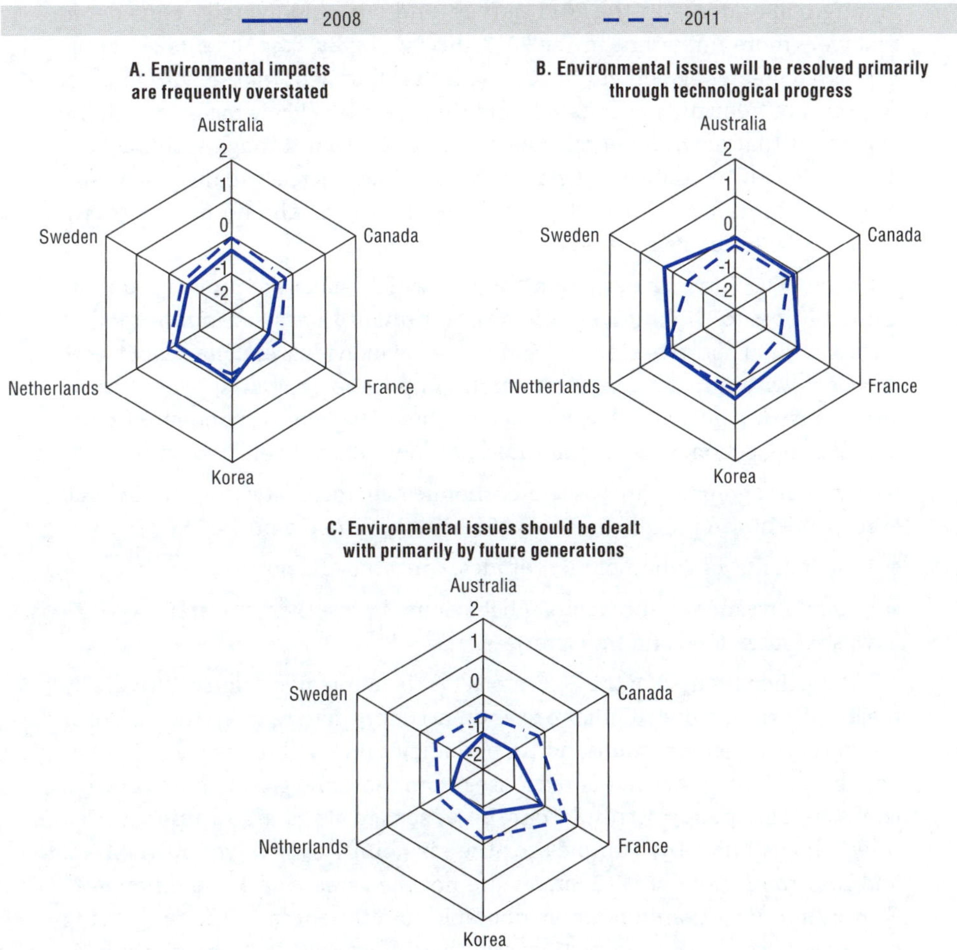

there were few environmental benefits appears as an important factor for not wanting to pay for "green" electricity. Looking at motivations to conserve resources, the provision of information on potential environmental impacts is reported to have a greater effect for water than for energy use. Conversely, more information on the household's consumption is thought to have a bigger effect on energy use.

Comparing some questions for the six countries involved in the two rounds of the Survey shows that Respondents ranked economic concerns relatively higher in 2011 than in 2008 and the opposite is true of environmental issues. Looking at the perceived seriousness of various environmental concerns, the biggest changes are with respect to biodiversity and natural resource depletion, with the latter considered relatively more important among respondents in the 2011 survey, and with biodiversity considered relatively more important in the 2008 survey. Finally, respondents generally expressed greater agreement in 2011 with the statement that "environmental impacts are frequently overstated" and they were less likely to agree with the statement that environmental issues will be resolved through technological progress. In all six countries there was a significant increase in the percentage of respondents who felt that environmental issues should be dealt with primarily by future generations.

The EPIC Survey has three attributes which make the project a unique contribution to the analysis of the relationship between public policy, environmental norms and attitudes, and a set of individual and household-level decisions with important environmental implications. First, it uses a single survey instrument to collect commensurable data from respondents from a broad cross-section of OECD countries. Secondly, data are elicited on:

- The demographic and socio-economic characteristics of individual respondents and the households of which they are members.
- Respondents' environmental attitudes, concerns and norms.
- Capital investment and habitual behaviours in five thematic areas (energy, waste, water, food and transport).

A further strength of the EPIC project is the implementation of surveys by means of a similar questionnaire over time. Unfortunately, given the nature of the survey implementation and its periodicity (every three years), it is not possible to develop a panel of responses from the same group of households. Nonetheless, general trends can be assessed. It is also important to re-emphasise that the samples for the six countries involved in the 2008 and 2011 rounds of the EPIC Survey are not the same, and some differences across time may therefore be attributable to differences in the underlying characteristics of the samples. However, every effort is made to adopt a similar sampling strategy in subsequent rounds of the survey. As further rounds are undertaken, it will be interesting to see how attitudes, consumption choices and behaviour develop over time.

Greening Household Behaviour
Overview from the 2011 Survey

ANNEX A

OECD 2011 Survey: Questionnaire

OECD Survey on Environmental Policy and Individual Behaviour Change (EPIC)

Canadian edit master – English version

1. How would you define your status in your current primary residence?

1. Married or living as a couple (with or without children)
2. Living with parents or other relatives
3. Living alone
4. Living as a single parent
5. Sharing a house/flat with non-family members
6. Living in hostel type accommodation, e.g. university dormitory, army base -> **Close survey**

2. Thinking about household purchases, expenditures and bills (such as utility bills, grocery shopping, car and household appliance purchases), would you say that:

1. You are frequently involved in these decisions
2. You are sometimes involved in these decisions
3. You are never involved in these decisions -> **Close survey**

Part A – SOCIO-DEMOGRAPHIC CHARACTERISTICS

3. Are you:

1. Male
2. Female

4. What year were you born?

5. How many people (including yourself) live in your household?

1. 1
2. 2
3. 3
4. 4
5. 5+

6. How many children (people under 18) live in your household?

1. 0
2. 1
3. 2
4. 3
5. 4
6. 5+

7. How many of these children are under 5 years old?

1. 0
2. 1
3. 2
4. 3
5. 4
6. 5+

8. Which of the following regions do you currently live in?

1. Alberta
2. British Columbia
3. Manitoba
4. New Brunswick
5. Newfoundland
6. Northwest Territories
7. Nova Scotia

8. Nunavut
9. Ontario
10. Prince Edward Island
11. Quebec
12. Saskatchewan
13. Yukon Territory

9. How many years of education did you complete after high school?

10. What is your current employment status?

1. Self-employed
2. Employee (full time, part time or on temporary leave)
3. Retired
4. Homemaker – househusband/wife
5. Seeking a job/unemployed
6. Student
7. Unable to work, e.g. disability
8. Other, please specify: ____________

11. How would you characterise your current occupation (or previous occupation if retired)?

1. Professional (e.g. medical, teacher, analyst)
2. Office, service or sales worker (e.g. police, clerical)
3. Trade worker or other technical occupation (e.g. plumber, computer technician)
4. Other manual worker (e.g. driver, labourer, cleaner)
5. Other, please specify: ____________

12. Please select the classification which most closely characterises the primary income earner's occupation (the person who earns the most)?

Please select the classification which most closely characterises the primary income earner's occupation (or previous occupation if retired):

0. I am the household's primary income earner (or our incomes are similar)
1. Professional (e.g. medical, teacher, analyst)
2. Office, service or sales worker (e.g. police, clerical)
3. Trade worker or other technical occupation (e.g. plumber, computer technician)
4. Other manual worker (e.g. driver, labourer, cleaner)
5. Other, please specify: ____________

13. What is your household's approximate annual income, after tax?

Please include income from everyone in your household from all sources, including wages, government pensions and benefits and investments:

1. USD 1-USD 24 200	=	(approx. USD 1-USD 2 000 monthly)
2. USD 24 201-USD 34 400	=	(approx. USD 2 001-USD 2 850 monthly)
3. USD 34 401-USD 41 800	=	(approx. USD 2 851-USD 3 500 monthly)
4. USD 41 801-USD 49 000	=	(approx. USD 3 501-USD 4 100 monthly)
5. USD 49 001-USD 56 700	=	(approx. USD 4 101-USD 4 700 monthly)
6. USD 56 701-USD 65 200	=	(approx. USD 4 701-USD 5 450 monthly)
7. USD 65 201-USD 75 200	=	(approx. USD 5 451-USD 6 250 monthly)
8. USD 75 201-USD 88 800	=	(approx. USD 6 251-USD 7 400 monthly)
9. USD 88 801-USD 127 000	=	(approx. USD 7 401-USD 10 600 monthly)
10. More than USD 127 000	=	(more than approx. USD 10 600 monthly)
11. Don't know		
12. Prefer not to answer		

14. How would you describe your household's current income?

1. Finding it very difficult to live on current income
2. Finding it difficult to live on current income
3. Coping on current income
4. Living comfortably on current income
5. Living very comfortably on current income

15. Do you and/or another member of your household own your current primary residence (with or without a mortgage)?

1. Yes
2. No

16. Is your primary residence:

1. An apartment in a building with fewer than 12 apartments in total
2. An apartment in a building with 12 or more apartments
3. A detached house
4. A semi-detached / terraced house
5. Other, please specify: ____________

17. What is the approximate size of your primary residence in square feet? (please estimate)

1. Less than 540 ft^2 (Less than 50 m^2)
2. 541 ft^2-1070 ft^2 (50 m^2-100 m^2)
3. 1071 ft^2-2150 ft^2 (101 m^2-200 m^2)
4. More than 2150 ft^2 (More than 200 m^2)
5. Don't know

18. How would you best describe the area in which you live?

1. Major town/city
2. Suburban (fringes of a major town/city)
3. Small town or village
4. Isolated dwelling (not in a town or village)

19. Approximately how many years have you lived in your primary residence?

1. Less than 2 years
2. 2 to 5 years
3. 6 to 15 years
4. More than 15 years

20. What is the postal code of your primary residence?

Part B – ATTITUDINAL CHARACTERISTICS

21. How satisfied are you with your life at the moment?

All things considered, please indicate how satisfied you feel with your life at the moment. The bottom of the ladder indicates dissatisfaction, while the top indicates high levels of satisfaction with your life. On which step of the ladder do you personally feel you stand at the moment, assuming that the higher up the step the more satisfied and the lower on the ladder the less satisfied you feel with your life?

Very dissatisfied										Very satisfied
0	1	2	3	4	5	6	7	8	9	10

22. In your view, what are the most serious issues facing the world today?

Please rank the following issues in order of their importance.

1 stands for the most important and 6 for the least important.

Drag or double click on an issue on the left to move it to the right hand side. If you want to reorder an issue once it is on the right-hand side, select it and then use the up and down arrows:

1. International tensions (e.g. terrorism, war)
2. Economic concerns (e.g. unemployment, inflation, financial crisis)
3. Environmental concerns (e.g. pollution, waste, climate change)
4. Health concerns (e.g. cancer, infectious disease)
5. Social issues (e.g. poverty, discrimination)
6. Personal safety (e.g. crime, theft)

23. How serious are the following environmental issues facing the world?

Please select one answer per row:

Not at all serious										Extremely serious	Don't know/ no opinion
0	1	2	3	4	5	6	7	8	9	10	❑

Waste generation
Air pollution
Climate change (global warming)
Water pollution
Natural resource depletion (forest, water, energy)
Endangered species and biodiversity

24. Have you voted in any of the following types of elections in the past 6 years?

Please select all that apply:

1. National/general elections
2. Municipal/local elections
3. None of the above

25. In the past 24 months, have you supported or participated in activities of any of the following types of groups/organisations? (This includes membership/subscription, personal time, and/or financial donations)

Please select as applies:

1. Parent-teacher association
2. Environmental organisation
3. Local community association
4. Charitable organisation (e.g. health, development, poverty)
5. Other association/organisation
6. None of the above

26. To what extent do you agree with each of the following statements?

Please select one answer per row:

	Strongly disagree	Disagree	Agree	Strongly agree	No opinion
I am not willing to do anything about the environment if others don't do the same					
Environmental impacts are frequently overstated					
Environmental issues should be dealt with primarily by future generations					
I am willing to make compromises in my current lifestyle for the benefit of the environment					
Policies introduced by the government to address environmental issues should not cost me extra money					
Environmental issues will be resolved in any case through technological progress					
Protecting the environment is a means of stimulating economic growth					

27. How trustworthy do you consider the following sources with regard to information on claims about the environmental impacts of products

Please select one answer per row:

Not at all trustworthy										Very trustworthy	No opinion
0	1	2	3	4	5	6	7	8	9	10	❑

1. Researchers, scientists and experts, e.g. at universities or research institutes
2. National/local governments
3. Environmental non-governmental organisations (NGOs)
4. Consumers' organisations
5. Manufacturers and retailers (including producer associations)

28. How satisfied are you with the following aspects of your local environment?

Please select one answer per row:

	Very dissatisfied	Dissatisfied	Satisfied	Very satisfied	No opinion
Air quality					
Water quality (lakes, rivers, sea)					
Access to green spaces (parks, forests)					
Level of noise					
Litter and rubbish in your area					

29. For each statement below, tick the box that comes closest to your opinion of how true it is.

Please select one answer per row:

	Probably true	Definitely true	Probably not true	Definitely not true	Don't know
a. Climate change is caused by a hole in the earth's atmosphere					
b. Every time we use coal, oil or gas we contribute to climate change					

30. Among the following logos/labels, please select the ones you recognise:

[Images are shown to the respondents].

❑ None of the above

31. For the logos/labels selected in 30, the following questions are asked:

	Yes	No	Don't know/ not applicable
Do you understand what this label means?			
Do you trust this label?			
Do you use this label in your purchasing decisions?			

32. In your opinion, when labelling a product how useful would the following types of information be for you?

Please select one answer per row:

Not useful 0	1	2	3	4	5	6	7	8	9	Very useful 10	Don't know/ not applicable ❑

Whether the product can be recycled
Total amount of greenhouse gas emissions contributing to climate change (e.g. carbon footprint)
Sustainable management of natural resources (e.g. fisheries, forestry)
Whether the welfare of animals has been respected
Whether social objectives have been respected (e.g. fair trade)

Part C – WASTE

The following section covers waste and recycling.

33. How often is your mixed waste collected from your primary residence or from containers where you dispose of your waste?

This excludes waste sorted for recycling/composting

1. Every day
2. Not every day but more than once a week
3. Once a week
4. Less often than once a week (e.g. every second week)
5. No collection available in my area
6. Don't know

34. On average, how much mixed waste does your household generate each week?

This excludes waste sorted for recycling/composting

First, choose the size of bag:

Second, please indicate approximately the number of bags of mixed waste that you generate on average in a week:

0 bags	15 or more	Don't know
|------------------------------	------------------------------|	❑

35. How well informed do you feel you are with regard to what is recyclable in your area?

Not at all informed										Very well informed
0	1	2	3	4	5	6	7	8	9	10

36. What are the waste collection services available for recyclable materials in your area?

Select all that apply:

	Door-to-door collection	Drop-off centres/ containers	Bring back with refund (to the retailer/ manufacturer)	Bring back with no refund (to the retailer/ manufacturer)	No recycling service available	Don't know
Glass bottles/ containers						
Plastic bottles/ containers						
Aluminium, tin and steel cans						
Paper/ cardboard						
Food and garden waste						

37. Please indicate approximately what percentage of the following waste items that your household recycles

Please consider only those items that you are actually able to recycle in your area:

0% (i.e. nothing)	100%	Don't know/ not available
|---|		❑

1. Glass bottles/containers
2. Plastic bottles/containers
3. Aluminium, tin and steel cans
4. Paper/cardboard
5. Food and garden waste

38. How important are the following factors in motivating your household to recycle?

Please select one answer per row:

Not important 0	1	2	3	4	5	6	7	8	9	Very important 10	Don't know/ not applicable ❑

It is beneficial for the environment
Reduce waste charges or collect refunds
I think it is my civic duty
I want to be seen by others as a responsible citizen

39. How is your household charged for the collection of mixed waste in your primary residence?

Please select one:

1. **Flat fee** (e.g. lump sum included in property taxes, charges or rent)
2. According to **volume** (e.g. per bag, per container, bag tags, etc.)
3. According to **weight** (e.g. per kg, pound, etc.)
4. According to **frequency** of collection (e.g. how often the waste is collected)
5. According to **size of household or residence**
6. Other form of charging, please specify: _________
7. **Not charged**
8. Don't know

40. In general, how do you dispose of old electronic equipment?

Please select as many as apply:

1. Store at home/store indefinitely
2. Dispose of with mixed waste
3. Take to shop
4. Use specialised disposal service
5. Give old goods to charity
6. They are collected with my recyclables
7. Periodic collection of hard/durable waste items
8. Other, please specify: _________

41. How do you dispose of old/unused medicines?

Please select as many as apply:

1. Store at home/store indefinitely
2. Dispose of with mixed waste
3. Take back to pharmacy or medical clinic
4. Flush down toilets or drains
5. Do not have unused medicines
6. Other, please specify: _________

42. To what extent would you support the following government actions to reduce household waste generation?

Please select one answer per row:

Don't support										Strongly support	No opinion
0	1	2	3	4	5	6	7	8	9	10	❑

1. Charge for waste collection by volume or weight
2. Encourage manufacturers to reduce packaging of their products
3. Encourage people to buy products with less packaging
4. Send an annual summary to each household/apartment block of the volume of waste collected
5. Educate the public on the environmental impacts of waste

Part D – TRANSPORT

The following section covers personal transport.
In this section, when using the word "car" we also include vans and sport utility vehicles (SUV).

43. How many cars are owned or used regularly by your household (including company cars)?

Number of cars: _________

44. Are any of these cars?

1. Hybrid car (combined electric motor + diesel/petrol engine)
2. Pure electric car (only electric motor)
3. Flex-fuel car (allows higher biofuel blends than conventional cars)
4. None of the above

45. Please enter details concerning your electric, hybrid or flex-fuel car

	Model year	Approximate total kilometres driven by all household members in this vehicle (monthly ***or*** yearly)	Please provide your car's average fuel consumption (litres per 100 km)	Fuel type	Is this the car used most often in your household?
Electric, Hybrid or Flex-fuel car	1. Pre-2000		1. Less or equal to 5L/100 km	1. Petrol/bioethanol	1. Yes
	2. 2000		2. 6 to 9L/100 km	2. Diesel/biodiesel	2. No
	3. 2001		3. 10 to 13L/100 km		
	4. 2002		4. 13 to 16L/100 km		
	5. 2003		5. More than 16L/100 km		
	6. 2004		6. Don't know		
	7. 2005				
	8. 2006				
	9. 2007				
	10. 2008				
	11. 2009				
	12. 2010				

46. Please enter the information concerning the car used most often in your household

	Fuel type	Approximate total kilometres driven by all household members in this vehicle (monthly ***or*** yearly)	Model year	Age of the car (years)	Please provide your car's average fuel consumption (litres per 100 km)
Car used most often	Unleaded petrol		1. Pre-2000	Less than 1 year old	Less or equal to 5L/100 km
	Diesel		2. 2000	1 to 5 years old	6 to 9L/100 km
	Liquefied / compressed natural gas / LPG		3. 2001	6 to 10 years old	10 to 13L/100 km
	Bioethanol (E20 to E100 – this fuel is NOT used in conventional cars)		4. 2002	11 to 15 years old	13 to 16L/100 km
	Bio-diesel (B20 to B100 – this fuel is NOT used in conventional cars)		5. 2003	More than 16 years old	More than 16L/ 100 km
	Biogas		6. 2004	Don't know	Don't know
	Leaded petrol		7. 2005		
	Diesel/petrol used for *hybrid cars ONLY*		8. 2006		
	Grid electricity (pure *electric cars only*)		9. 2007		
	Don't know		10. 2008		
			11. 2009		
			12. 2010		

47. How many motorcycles are owned or used regularly by your household?

Number of motorcycles: _________

48. How important are the following reasons for your household not having a car?

Please rate the level of importance of the following factors:

Not important										Very important	Don't know/ not applicable
0	1	2	3	4	5	6	7	8	9	10	❑

1. Can't afford a car
2. From our place of residence a car is not necessary/Do not need a car
3. No one can drive (e.g. no drivers licence, disability)
4. Environmental concerns
5. Road congestion/traffic jams
6. Too few parking spaces
7. Use car share/hire car when necessary

49. How important are the following factors in your choice of car?

Please select one answer per row:

Not important 0	1	2	3	4	5	6	7	8	9	Very important 10

1. Price
2. Fuel consumption
3. Environmental impacts
4. Comfort
5. Safety
6. Performance and handling
7. Reliability
8. Brand affinity

50. Have you purchased a car in the last 5 years?

1. Yes
2. No

51. When you bought this car, did you:

a) Pay less money because when you purchased your vehicle you returned an old car as part of a government vehicle scrappage programme?

b) Pay less money because you purchased a fuel-efficient and/or low-polluting vehicle? (reduced tax/registration fee or provision of rebate)

52. How much more would you pay for an electric car (only electric motor) compared to a conventional car?

0% (i.e. same price)	100% or more (i.e. twice as much or more)	Don't know	Not applicable
|------------------------------	------------------------------|	❑	❑

53. Why are you not willing to pay more for an electric car? [Asked if 52 = 0%]

Please indicate how much you agree or disagree with the following statements.

Please select one answer per row:

	Strongly disagree	Disagree	Agree	Strongly agree	No opinion
If I had more money I would pay more for an electric car					
I prefer conventional models to the electric cars available					
Adequate infrastructure is not yet available for electric cars (i.e. limited charging facilities)					
Inconvenience of electric cars due to the need to charge frequently					
I don't think there are environmental benefits of electric cars					
I don't think I should have to pay extra for an electric car					

54. What means of transport would you use to get to the public transport stop/station which is most convenient for your daily commute?

1. By car
2. By foot
3. By bicycle
4. Not applicable (e.g. I don't commute)

55. Using this means, how long does it take to get to the most convenient public transport stop/station?

1. Less than 5 minutes
2. 5-15 minutes
3. 16-30 minutes
4. Over 30 minutes
5. Don't know
6. No public transport/station available
7. Not applicable

56. How many kilometres do you personally drive (car/motorcycle) during a typical week?

1. Do not drive
2. Less than 25 km
3. 26-100 km
4. 101-250 km
5. 251-500 km
6. More than 500 km
7. Don't know

57. Please tick if this includes kilometres driven for professional activities such as visiting patients/clients.

58. How important are the following factors in encouraging you to drive (car/motorcycle) less?

Not important										Very important	Don't know/ not applicable
0	1	2	3	4	5	6	7	8	9	10	❑

1. Increased cost of car/motorcycle use (e.g. fuel, parking and tolls)
2. Improved public transport (e.g. better or cheaper)
3. More and safer cycling paths
4. Fewer parking spaces
5. More information on the environmental impacts of driving relative to other means of transport

59. Would any of the following aspects of public transport encourage you to use your car/motorcycle less?

Please select all that apply:

1. More convenient (e.g. stops closer to home and destination)
2. More reliable (e.g. fewer delays, strikes)
3. More rapid (e.g. higher frequency, speed)
4. More comfortable (e.g. less crowded)
5. More secure (e.g. improved personal safety)
6. More affordable

❑ None of the above

60. What is your main mode of transportation for each of the following activities?

Please choose the mode which accounts for the greatest distance:

	Walking	Car/ motorcycle/ taxi	Public transport	Bicycle	Not applicable
Daily commute to and from work					
Food shopping					

61. Approximately how long does it take you to get to work (one way)?

1. Less than 15 mins
2. 15-30 mins
3. 31-45 mins
4. 46 mins-1 hour
5. More than 1 hour

62. Compared to your usual mode of transport, how long would it take to get to work using these different modes of transport?

	Car/motorcycle	Public transport	Bicycle	Walking
More than 30 minutes ***shorter***				
16-30 minutes ***shorter***				
5-15 minutes ***shorter***				
Same time (approximately)				
5-15 minutes ***longer***				
16-30 minutes ***longer***				
More than 30 minutes ***longer***				
Don't know / Not possible				

63. During the past year, have you done any of the following?

Select all that apply:

1. Used car pooling
2. Used a car-sharing scheme (i.e. short-term rental)
3. Used low rolling resistance tyres
4. Offset your carbon emissions from flights
5. Adapted your driving style to use less fuel (e.g. reduce speed, reduce air-conditioning use)

❑ None of the above

64. To what extent would you support the following government actions to reduce motor vehicle CO_2 emissions?

Please select one answer per row:

Don't support 0	1	2	3	4	5	6	7	8	9	Strongly support 10	No opinion ❑

1. Stricter limits on vehicle fuel efficiency
2. Higher taxes on automotive fuels
3. A price bonus (or tax credit) for purchasing a less-polluting car
4. Invest in public transport infrastructure (e.g. buses, metro, bicycle lanes)
5. Educate the public of the environmental impacts of private transport
6. Label vehicles according to their environmental impact

Part E – ENERGY

The following section covers residential energy use. In order to save you time, it would be useful to have your electricity bills at hand.

65. In your household, do you pay for your electricity according to how much electricity you use? (e.g. individual electricity metering)

1. Yes
2. No
3. Don't know

66. Approximately how much was the total annual cost for electricity consumption for your primary residence in the last year?

Please indicate if possible amount in USD and corresponding annual consumption:

Amount in USD per year *Please provide answer to the nearest dollar*	Quantity of electricity consumed in kWh

❑ Don't know

67. Which of the following energy sources do you use for space heating/cooling, water heating and cooking?

Choose as many as apply:

Space heating/cooling	Water heating	Cooking
1. Electricity	1. Electricity	1. Electricity
2. Gas, oil, coal and other fossil fuels	2. Gas, oil, coal and other non-renewable	2. Gas (bottled, main, etc.)
3. Wood or burning pellets	3. Thermal solar panel	3. Other, please specify_____
4. District heating	4. Other, please specify_____	4. Don't know
5. Ground-source heat pump	5. Don't know	
6. Other, please specify_____		
7. Don't know		

68. Has your electricity provider proposed the following services to your household?

Differentiated electricity rate for peak time (e.g. early evening) and off-peak time (e.g. night)
Smart electricity meters allowing you to monitor consumption by viewing electricity usage in real time
A "renewable/green" energy tariff where you are guaranteed a specified amount of renewable electricity in your supply *By renewable energy we mean energy sources such as wind, solar, geothermal, hydro*

1. Yes and I have chosen this option
2. Yes, but I have not chosen this option
3. No, and I am not interested
4. No, but I would be interested
5. Don't Know

69. Do you use the information collected by your smart meter?

1. The smart meter has helped me to reduce my household's electricity consumption
2. I use the smart meter but I have not reduced consumption
3. I don't use the information as I don't have the time or motivation
4. Other, please specify: ________________

70. If a renewable/green energy tariff were available with another electricity provider at no extra financial cost, would you be willing to change provider?

1. Yes, I would definitely change provider
2. I would probably change provider
3. No, I would not change provider

71. What is the maximum percentage increase on your annual bill you are willing to pay to use only renewable energy?

Please assume that your energy consumption remains constant:

0% (i.e. same price)	100% or more (i.e. twice as much or more)	Don't know
|--	--------|	❑

72. Why would you not be willing to pay more to use only renewable energy? [Asked if 71 = 0%]

Please indicate how much you agree with the following statements.

Please select one answer per row:

	Strongly disagree	Disagree	Agree	Strongly agree	No opinion
If I had more money I would pay more to use only renewable energy					
I consider there is already enough renewable energy in the general electricity supply mix					
I do not trust that paying extra for renewable energy would actually increase renewables in the mix					
I do not believe there are environmental benefits associated with renewable energy					
I am not interested in renewable energy					
I don't think I should have to pay extra					

73. Did you take energy costs into account when purchasing or renting your current primary residence?

1. Yes
2. No
3. Not sure

74. How many of the following appliances do you have in your home?

1. Fridge or combined fridge-freezers
2. Separate freezers
3. Televisions
4. Electric air conditioners
5. Computers
6. Clothes dryers

75. Does your household have the following items?

	Yes	No	Don't know
Flat screen TV			
Mobile phone with connection to the internet			
Internet telephony (e.g. Skype voice calls)			

76. How often do you perform the following in your daily life?

	Never	Occasionally	Often	Always	Don't know / Not applicable
Turn off lights when leaving a room					
Cut down on heating/air conditioning to limit energy consumption					
Only run full loads when using washing machines or dishwashers					
Wash clothes using cold water (e.g. 30^0 C) rather than warm/hot water (over 30^0 C)					
Switch off standby mode of appliances/ electronic devices (TV, computer)					
Air dry laundry rather than using a clothes dryer					

77. Has your household installed any of the following items over the past ten years in your current primary residence?

Please select one answer per row:

	Yes	No	Already equipped	Not possible (not feasible in my house/ apartment OR my landlord would need to install this)
Top-rated energy-efficient appliances (e.g. top-rated washing machines, refrigerators)				
Low-energy light bulbs (compact fluorescent, LED)				
Energy-efficient windows (e.g. double or triple glazed windows)				
Thermal insulation of walls/roof				
Heat thermostats				
Solar panels for electricity or hot water				
Wind turbines				
Ground-source heat pumps (ground temperature is used to provide heating/cooling via a compressor and pipes buried under ground)				

78. For which of the following has your household benefited from government (or utility company) financial support (e.g. grants or preferential loans)? [For items selected as Yes in 77]

79. How important would the following factors be in encouraging you to reduce your energy consumption?

Please indicate the level of importance for each item:

Not important										Very important	Don't know
0	1	2	3	4	5	6	7	8	9	10	❑

More practical information on how to reduce energy consumption at home
Higher energy prices
More information on the environmental impacts of energy consumption
More information on the energy consumption of my household
Finding that my household uses more energy than similar households
Easier identification of energy-efficient appliances
Less expensive to invest in energy-efficient equipment

Part F – FOOD

80. How important are the following factors in your food shopping choices?

Not important 0	1	2	3	4	5	6	7	8	9	Very important 10	Don't know/ not applicable ❑

Price	
Health	
Environmental aspects	
Seasonal and locally-produced	
Freshness and taste	
Animal welfare	
Familiarity and preferred brands	

81. In a normal week where do you shop for food?

Please tick all that apply:

1. Out-of-town large supermarkets
2. Local supermarkets
3. Neighbourhood specialist food shops (e.g. bakeries, butcheries, fruit and vegetable shops)
4. Markets – street or specialty markets
5. Convenience stores
6. Internet shopping
7. Other

82. To what extent do you agree with the following statements?

Please select one answer per row:

	Strongly disagree	Disagree	Agree	Strongly agree	No opinion
Consuming meat and other animal products has significant negative environmental consequences					
Importing food from distant areas has significant negative environmental consequences					
Food waste has significant negative environmental consequences					

83. Does your household usually?

	Yes	No	Not applicable
Compost your food waste			
Choose food items with less packaging			
Eat food that is in season and locally grown			
Limit or avoid consumption of meat			
Use reusable shopping bags for food shopping			
Choose fish certified as sustainable over other types of fish			

84. Approximately what proportion of food bought by your household is thrown away?

Please exclude non-edible parts of food, e.g. peelings, apple cores, etc.:

0% (i.e. none) |---| 100% Don't know ❑

85. Please estimate the percentage of your household's food expenditures for meat and poultry which are labelled as taking animal welfare into account:

0% (i.e. none) |-----------------------------------| 100% Don't know ❑ Not applicable ❑

86. What is the maximum percentage price increase you are willing to pay for meat and poultry which take animal welfare into account compared to conventional substitutes?

0% (e.g. same price) |-----------------------| 100% (e.g. twice as much) |-----------------------| 200% or more (e.g. three times as much or more)

❑ Don't know

❑ Not applicable

> The following section covers organic food consumption.
>
> By organic we mean a production process where, depending on the standard, fewer chemicals (i.e. pesticides, fertilizers, drugs, additives), if any, are used.

87. Please estimate the percentage of your household's food expenditures for fresh fruit and vegetables which are labelled as being organic:

0% (i.e. none)	100%	Don't know/ not applicable
|--	------------------------------------|	❑

88. What is the maximum percentage price increase you are willing to pay for organic fresh fruit and vegetables compared to conventional substitutes?

0% (e.g. same price)	100% (e.g. twice as much)	200% or more (e.g. three times as much or more)	Don't know/ not applicable
|------------------------------	|--------------------------------	|	❑

89. Why are you not willing to pay extra for fresh fruit and vegetables? [Asked if 88 = 0%]

Please indicate how much you agree or disagree with the following statements.

Please select one answer per row:

	Strongly disagree	Disagree	Agree	Strongly agree	No opinion
If I had more money I would pay more for organic food					
I do not trust the food is actually grown organically					
I do not believe organic products are better for health or the environment					
I do not think I should have to pay extra					

90. How important would the following factors be in encouraging you to increase your consumption of organic food?

Please indicate how important this factor is to you:

Not important 0	1	2	3	4	5	6	7	8	9	Very important 10	Don't know/ not applicable ❑

Better availability of organic products
Lower price of organic products
Better appearance of the food
Easier identification of organic products
More trust in health benefits of organic products
More trust in environmental benefits of organic products
More trust in certification and labelling of organic products

Part G – WATER

The following section covers water consumption and use.

91. In your household, do you pay for water according to how much water you use?

1. Yes (e.g. individual water metering)
2. No (e.g. not charged or charged a flat fee such as lump sum included in charges, property taxes or rent)
3. Other (e.g. charged communally for water with other households in my building)
4. Don't know

92. How often do you do the following in your daily life?

Please select one answer per row:

	Never	Occasionally	Often	Always	Not applicable
Turn off the water while brushing teeth					
Plug the sink when washing the dishes by hand					
Water garden in the coolest part of the day to reduce evaporation and save water					
Collect rainwater (e.g. in water tanks) or recycle waste water					
Rinse dishes before putting them in the dishwasher					
Take showers instead of baths					

93. Did you take water efficiency into account when you last purchased either a washing machine or a dishwasher?

1. Yes
2. No
3. Don't know
4. Not applicable

94. Has your household invested in the following appliances/ devices in the past 10 years in your current primary residence?

	Yes	No	Already equipped	Not possible (not feasible in my house/apartment OR my landlord would need to install this)
Low-volume or dual-flush toilets				
Water flow-restrictor taps / low-flow shower head				
Water tank to collect rainwater				

95. For which of the following has your household benefited from government (or utility company) financial support to make this investment (e.g. grants, incentives or free goods)? [For items selected as Yes in 94]

Please select all that apply.

96. How important would the following factors be in encouraging you to reduce your household's water consumption?

Please indicate the level of importance to you:

Not important 0	1	2	3	4	5	6	7	8	9	Very important 10	Don't know/ not applicable ❑

More practical information on things you can do to save water at home
Higher water prices
More information on the environmental impacts of water consumption
More information on the water consumption of my household
Finding that my household uses more water than similar households
Easier identification of water-efficient appliances
Less expensive to invest in water-efficient equipment

97. For your normal household drinking water which of the following do you usually drink?

1. Straight from the tap
2. Purified/filtered/boiled tap water
3. Bottled mainly still/flat
4. Bottled mainly sparkling
5. Natural source (e.g. rainwater/surface water/well)
6. Other, please specify:______________

98. How satisfied are you with the following aspects of your tap water?

Not satisfied 0	1	2	3	4	5	6	7	8	9	Very satisfied 10	Don't know ❑

1. Taste
2. Health impacts

ANNEX B

OECD 2011 Survey: Implementation

About the EPIC Survey

OECD work on Greening Household Behaviour involves implementing a periodic survey on Environmental Policy and Individual Behaviour Change (EPIC) across a number of countries and areas. This activity was initiated with the first survey in 2008 and the second round was implemented in early 2011 in eleven countries: Australia, Canada, Chile, France, Israel, Korea, Japan, the Netherlands, Spain, Sweden and Switzerland. Details on the project timeline are provided in Table B.1.

Table B.1. **Project timeline**

Activity	Timeframe
OECD questionnaire design, with inputs from the advisory committee meeting	Mar. 2010-Dec. 2010
Selection of survey provider, call for tender (July 2010)	July 2010-Sept. 2010
Pilot: 500 respondents	Nov. 2010
Translations: 14 different versions	Dec. 2010
Creation and testing of online questionnaires	Dec. 2010-Feb. 2010
Soft launch	Feb. 2010
Full implementation of the OECD EPIC Survey, simultaneously in 11 countries	Feb.-Mar. 2011

In 2011, approximately 1 000 households were surveyed in each of the eleven countries using an internet-based questionnaire resulting in a total sample of more than 12 200 households (Table B.2). As in 2008, information was collected on household characteristics (e.g. age, income, education), environmental attitudes (e.g. concerns for the environment), and the use of eco-labelling and household behaviours in five key areas: residential energy use, waste generation and recycling, food consumption, personal transport choices, and water consumption.

The statistical data for Israel are supplied by and under the responsibility of the relevant Israeli authorities. The use of such data by the OECD is without prejudice to the status of the Golan Heights, East Jerusalem and Israeli settlements in the West Bank under the terms of international law.

Selection of service provider

The OECD ran a *Call for Tender* to select a survey service provider specialised in the implementation of large international web-surveys using online consumer panels in different countries. As in 2008, an internet-based survey was the OECD's preference because of ease of implementation across multiple OECD countries and cost. The underlying aim was to obtain a representative sample with the assessment criteria focusing on panel size, panel quotas, panel recruitment and management, online survey design abilities, responsiveness and cost. The OECD selected Global Market Insite (GMI), to run the EPIC Survey. The tasks included hosting and programming the online questionnaire, selecting a nationally representative sample of respondents from established online panel members, and collecting and cleaning the data.

Table B.2. **Number of respondents per country**

Australia	996
Canada	1 122
Chile	1 027
France	1 227
Israel	1 168
Japan	1 043
Korea	1 116
Netherlands	1 301
Spain	1 101
Sweden	1 012
Switzerland	1 089
Total	**12 202**

Questionnaire design and pilot testing

The OECD EPIC Survey questionnaire is composed of seven parts: two parts dealing with socio-demographic and attitudinal characteristics, and five thematic parts relating to household behaviour in the five environmental areas of interest: waste generation and recycling, personal transport choices, residential energy use, food consumption and residential water use.

The questionnaire is composed of approximately 90 closed questions with a combination of binary, class and Likert scale questions. As filter questions were used, respondents were not required to answer all questions.

The questionnaire was developed with the inputs from an advisory committee set up to inform the project and composed of government representatives from the countries involved in the survey and other experts, including from the OECD directorates working in related areas (Environment, Trade and Agriculture; Science, Technology and Investment), and the

International Energy Agency. The 2011 questionnaire was largely based on that used in the 2008 round, with improvements based on knowledge gained from the first round, and new areas were explored: adoption of eco-innovation, knowledge questions, policy preferences and country-specific questions.

A meeting of the advisory committee was held in Paris in June 2010 with delegates from participating countries and other experts to discuss the 2011 questionnaire. The draft questionnaire to be tested in pilot surveys was finalised to reflect the outcome of the meeting and inputs provided by members of the advisory committee, bearing in mind the constraints in terms of total length of the questionnaire.

A pilot survey of 500 respondents run in November 2010 was used to refine the questionnaire. The pilot, run in three English-speaking countries (Australia, Canada, Israel), allowed respondents to flag difficult questions, and the survey times (per question) were analysed. In addition, the data were also checked for variation and consistency, and a significant number of changes were introduced on the basis of the findings.

Survey technology and translation

The online questionnaires were programmed as an internet application, usable in standard web browsers. The technology permitted a variety of response formats for different questions. For example, respondents could use a sliding bar to indicate how likely they thought specific events were, or how much they would support a given policy. Annex B provides links to the online questionnaires for each country (see below), as well as individual thematic sections of the Canadian English questionnaire (as an example).

The target median completion time for the online survey was approximately 30 minutes, an objective which was largely met (see below). The survey was visually appealing and offered a number of question types, for instance: different attributes could be ranked by displacing them on screen; eco-labels were displayed and respondents could select those they recognised; and sliders were used for willingness-to-pay questions, thereby reducing the influence of framing when compared to a conventional questionnaire.

Once the online survey was programmed in English, the survey was translated into Dutch, French, German, Hebrew, Japanese, Korean, Spanish and Swedish with 14 distinct versions created once country-specific vocabulary was taken into account. Translations complied with the International Standard ISO 20252: all translations were carried out and double-checked by native speakers. Members of the advisory committee conducted final checks on the translated questionnaires.

Respondent targeting, recruitment and quota sampling

The target respondent for the survey was someone who was between 18 and 70 years of age and who had some influence on household purchasing decisions, expenditures and bills (such as utility bills, car and appliances purchases). Respondents were recruited from GMI's in-country panels. In some countries, GMI partnered with in-country firms with their own panels in order to further increase panel size. All partners were selected on the basis of quality of panel management. Specifically, GMI and its partners managed their respondent panels in adherence to the ESOMAR 26, which is a standard for transparency and accountability in the use of respondent panels for web-based survey research.

To ensure representativity in the sample, quotas were set for age, gender, region and income (see section below for how quota targets were set). When quotas were filled, respondents with these characteristics were stopped from completing the questionnaire. Panellists selected on the basis of these characteristics received e-mails inviting them to respond to the survey. No mention was made of the topics addressed in the questionnaire. To promote participation in the survey, a small in-kind incentive, worth approximately 5 to 10 US dollars was offered to respondents. To limit the risk of recruiting "professional respondents", GMI only permitted panellists to answer and receive compensation for, up to five questionnaires per year.

Potential respondents who started the questionnaire were asked whether they met the screening criteria (living in non-institutional settings and influential in household financial decisions). If they did not meet the criteria, they were thanked for their time and screened out of the sample.

Despite rigorous efforts at stratification and quota sampling, it is important to acknowledge that there may be some respondent characteristics that were not observed and which correlate with internet use. This correlation of unobserved characteristics could introduce a selection bias in the sample. It is therefore recommended that researchers drawing conclusions from these data carefully consider how this selection bias based on internet use could affect their results.

Defining the quotas and setting the targets

Statistics from each country's most recent census, provided by the national statistical agencies, were used to establish the quota targets. Age was stratified using the following groups: 18 to 4, 25 to 34, 35 to 44, 45 to 54 and 55 to 69.[1] Gender was approximately half male and half female for all countries. Region was stratified and quotas created using three to five regions. However, invitations were targeted according to a larger number of sub-regions within the regions. For Switzerland, quota targets were set for nationally representative proportions of French- and German-speakers, rather than for regions. For income stratification, households' after-tax income quintiles were estimated for each country, then responses from the survey income question were used to fill the quotas. Income quotas were not used in Israel and Chile.

Towards the end of survey implementation, the distribution of respondents was checked to see how close they were to target quotas. In practice, every effort was made to ensure that quotas were within 10% of the target and further targeted e-mails were sent, soliciting respondents in underfilled quotas. However, in a small number of cases, it was not possible to meet this requirement. Details on achievement of the quota targets are given below.

Response times and drop-out rates

Table B.3 shows the drop-out rates by country. The drop-out rate is calculated as the fraction of respondents who started the questionnaire but then dropped out. Potential respondents who were removed because of quotas or were screened out of the questionnaire after filter questions (e.g. did the respondent take decisions about household finances?) are not included in the calculation.[2] The overall drop-out rate was 21%, but this varied from 13% in Korea to 35% in Chile.

Table B.3. **Screened, drop-outs and completions, by country**

	Screened out		Stopped questionnaire		Completed questionnaire	Drop-out rate[1]
	Quotas	Filter questions	**At** filter questions	**After** filter questions		
Total	**18 122**	**2 026**	**6 397**	**3 199**	**12 303**	**20.6%**
Australia	641	95	340	281	1 007	21.8%
Canada	479	100	327	270	1 130	19.3%
Chile	366	295	588	550	1 034	34.7%
France	2 147	138	530	406	1 234	24.8%
Israel	1 475	169	819	270	1 177	18.7%
Japan	4 759	546	1 887	312	1 047	23.0%
Korea	4 311	303	562	175	1 134	13.4%
Netherlands	1 965	116	488	357	1 310	21.4%
Spain	1 081	111	336	197	1 108	15.1%
Sweden	829	101	308	202	1 030	16.4%
Switzerland	69	52	212	179	1 092	14.1%

1. The drop-out rate is calculated as: (Stopped **after** filter questions)/(Stopped **after** filter questions + Completed).

Median response times by country varied from 28 minutes (Korea) to 41 minutes (Chile), with the medians in all countries being less than 35 minutes except Chile (see Table B.4). The overall median is just over 31 minutes. The mean response times are much higher, but this is due to a small number of observations per country in which the respondent left the software open for a period of time, returning much later to complete the survey.

In cleaning the data, it was observed that a small fraction of respondents were "speeders", i.e. these individuals progressed through the survey at such a rapid pace that the validity of their responses is suspect. To address this issue, respondents who completed the questionnaire in less than half of the median response time for their country were removed from the dataset. The number of cases removed in this way is summarised in the Table B.4.

Table B.4. **Questionnaire completion time and speeders**[1]

	Completion time (min)		Speeders	
	Median	Mean	Number	Frequency
Total	**31.6**	**48.4**	**654**	**5.4%**
Australia	30.4	58.6	67	6.7%
Canada	32.1	50.9	74	6.6%
Chile	40.9	54.5	19	1.9%
France	29.8	51.1	75	6.1%
Israel	34.2	61.8	61	5.2%
Japan	28.0	37.0	61	5.8%
Korea	27.8	36.9	49	4.4%
Netherlands	27.8	36.9	41	3.2%
Spain	32.8	46.9	90	8.2%
Sweden	32.0	49.5	52	5.1%
Switzerland	32.5	47.5	65	6.0%

1. Speeders are defined as those completing the questionnaire in less than half of the median completion time for their country.

Links to the EPIC online questionnaires

The links below give you access to the full version of the OECD questionnaires implemented online in early 2011 in eleven countries.

Please note that these links are archived and that your replies will not be registered. To review the links successfully, each time you want to test the survey you have to follow some guidelines: before, open your Internet explorer, click on "Tools", select "Internet options" then select "Browsing history" and delete "Temporary Internet files" and "Cookies".

Country	Link to online questionnaire
Australia	*http://qsurvey.gmisurveys.com/dc/index.html?p=Pny92y2*
Canada (English)	*http://qsurvey.gmisurveys.com/dc/index.html?p=PWuYTGX*
Canada (French)	*http://qsurvey.gmisurveys.com/dc/index.html?p=jwgS11*
Chile	*http://qsurvey.gmisurveys.com/dc/index.html?p=jwoQQP*
France	*http://qsurvey.gmisurveys.com/dc/index.html?p=vtWmd9*
Israel (English)	*http://qsurvey.gmisurveys.com/dc/index.html?p=1jCsEO*
Israeli (Hebrew)	*http://qsurvey.gmisurveys.com/dc/index.html?p=eKOojd*
Japan	*http://qsurvey.gmisurveys.com/dc/index.html?p=Pwpt9Xa*
Korea	*http://qsurvey.gmisurveys.com/dc/index.html?p=jwgS1c*
Netherlands	*http://qsurvey.gmisurveys.com/dc/index.html?p=jwoQQg*
Spain	*http://qsurvey.gmisurveys.com/dc/index.html?p=1jCsEh*
Sweden	*http://qsurvey.gmisurveys.com/dc/index.html?p=vtWmdU*
Swiss (French)	*http://qsurvey.gmisurveys.com/dc/index.html?p=eKOojj*
Swiss (German)	*http://qsurvey.gmisurveys.com/dc/index.html?p=Pf25gRk*

The links below give access to the specific thematic sections of the English questionnaire (Canadian version).

Thematic areas	Link to online sections
Waste generation and recycling	*http://qsurvey.gmisurveys.com/dc/index.html?p=PWuYTGX&test=1*
Personal transportation	*http://qsurvey.gmisurveys.com/dc/index.html?p=PWuYTGX&test=2*
Residential energy use	*http://qsurvey.gmisurveys.com/dc/index.html?p=PWuYTGX&test=3*
Food consumption	*http://qsurvey.gmisurveys.com/dc/index.html?p=PWuYTGX&test=4*
Water use	*http://qsurvey.gmisurveys.com/dc/index.html?p=PWuYTGX&test=5*

Quota targets relative to samples, by country

As mentioned in the main text, each country's most recent statistics, as provided by national statistical agencies, were used for setting the quota targets. The variables for which quotas were set were: *a)* gender, *b)* age group, *c)* region and *d)* household income. Some country-specific modifications to this standard procedure are discussed below for Korea, Chile, Israel and Switzerland. These are indicated in grey in the table below.

For **Korea**, the age quota for the oldest group was modified so that only respondents under 65 were targeted (because of patterns of internet use in that country). However, as shown in the table below, there was significant difficulty in achieving the target for the oldest age group in this country. Therefore, in order to obtain a better representation of the elderly population in Korea, respondents over 55 years of age were included in the sample (provided they were not eliminated on the basis of other quota targets or screening questions).

For **Chile** and **Israel**, the samples were not stratified by income, owing to the limited availability of reliably estimated income distributions for these countries. However, in order to gauge how well the samples represented the economic profile of the general population for these countries, we included extrapolated income quintiles for these two countries. These quintiles were obtained by using estimates of mean income levels (for centring the income distribution) and the Gini coefficients for these two countries (for setting the spread of the income distribution).

For **Switzerland**, rather than setting quota targets for regions, quota targets were set for language: French and German.

Notes

1. The oldest age quota in Korea was set at 65 rather than 69 because of the composition of the web-using population in that country.
2. See Annex A, for the full 2011 EPIC Survey questionnaire.

Table B.5. **Quota targets relative to samples**

AUSTRALIA	Target	Sample	+/-20%
Gender			
Male	49.5%	49.2%	
Female	50.5%	50.8%	
Age			
18-24	14.2%	15.9%	
25-34	20.3%	18.7%	
35-44	22.3%	23.2%	
45-54	20.9%	21.0%	
55-69	22.3%	21.2%	
Household income			
AUD 0-36 300	20.0%	20.9%	
AUD 36 301-52 400	20.0%	17.3%	
AUD 52 401-69 900	20.0%	16.2%	
AUD 69 901-94 500	20.0%	17.0%	
Over AUD 94 500	20.0%	17.3%	
Don't know		3.3%	
Prefer not to answer		8.0%	
Region			
New South Wales	32.8%	30.3%	
Victoria	24.7%	25.6%	
Queensland	19.9%	20.6%	
South Australia + Western Australia + Tasmania + Northern Territory + Australian Capital Territory	22.6%	23.5%	

KOREA	Target	Sample	+/-20%
Gender			
Male	50.8%	49.8%	
Female	49.2%	50.2%	
Age			
18-24	14.7%	17.3%	
25-34	24.7%	27.9%	
35-44	25.4%	28.1%	
45-54	22.3%	21.4%	
55-69	12.8%	5.2%	*
Household income			
KRW 0-19 800 000	20.0%	14.2%	*
KRW 19 800 001-29 600 000	20.0%	17.1%	
FRW 29 600 001-39 000 000	20.0%	19.3%	
KRW 39 000 001-53 000 000	20.0%	20.9%	
Over KRW 53 000 000	20.0%	22.3%	
Don't know		4.4%	
Prefer not to answer		1.8%	
Region			
Seoul	21.4%	25.1%	
East	31.1%	18.0%	*
North	26.8%	36.9%	*
South	20.7%	20.0%	

NETHERLANDS	Target	Sample	+/- 20%
Gender			
Male	50.2%	50.1%	
Female	49.8%	49.9%	
Age			
18-24	12.3%	11.2%	
25-34	18.0%	14.9%	
35-44	23.1%	23.4%	
45-54	21.4%	23.9%	
55-69	25.2%	26.6%	
Household income			
EUR 0-19 800 000	20.0%	15.7%	*
EUR 19 800 001-29 600 000	20.0%	16.1%	
EUR 29 600 001-39 000 000	20.0%	14.7%	*
EUR 39 000 001-53 000 000	20.0%	11.3%	*
Over EUR 53 000 000	20.0%	6.1%	*
Don't know		10.4%	
Prefer not to answer		25.9%	
Region			
Noord-Nederland	10.4%	10.3%	
Oost-Nederland	21.2%	21.6%	
West-Nederland	46.7%	47.4%	
Zuid-Nederland	21.7%	20.8%	

SPAIN	Target	Sample	+/- 20%
Gender			
Male	50.4%	50.8%	
Female	49.6%	49.2%	
Age			
18-24	11.4%	12.5%	
25-34	23.8%	22.6%	
35-44	23.7%	25.2%	
45-54	19.6%	21.1%	
55-69	21.5%	18.6%	
Household income			
EUR 0-16 000	20.0%	18.2%	
EUR 16 001-23 400	20.0%	18.2%	
EUR 23 401-31 700	20.0%	17.3%	
EUR 31 701-44 000	20.0%	15.7%	*
Over EUR 44 000	20.0%	14.1%	*
Don't know		3.3%	
Prefer not to answer		13.2%	
Region			
North	23.6%	23.6%	
East	33.1%	33.2%	
South	29.0%	28.7%	
Madrid metropolitan	14.3%	14.5%	

SWITZERLAND	Target	Sample	+/- 20%
Gender			
Male	50.0%	*47.7%*	
Female	50.0%	52.3%	
Age			
18-24	12.0%	*12.6%*	
25-34	19.0%	*18.8%*	
35-44	23.5%	*21.4%*	
45-54	21.2%	21.2%	
55-69	24.3%	26.1%	
Household income			
CHF 0-56 700	20.0%	*25.0%*	*
CHF 56 701-76 200	20.0%	*19.2%*	
CHF 76 201-96 100	20.0%	*16.1%*	
CHF 96 101-126 700	20.0%	17.0%	
Over CHF 126 700	20.0%	9.0%	*
Don't know		2.0%	
Prefer not to answer		11.6%	
Region (not used)			

ISRAEL	Target	Sample	+/- 20%
Gender			
Male	50.0%	45.6%	
Female	50.0%	54.4%	
Age			
18-24	18.0%	19.5%	
25-34	25.0%	29.0%	
35-44	21.0%	21.1%	
45-54	17.0%	15.7%	
55-69	19.0%	14.7%	*
Household income* (not used)			
ILS 0-86 200	20.0%	25.3%	*
ILS 86 201-127 700	20.0%	19.4%	
ILS 127 701-163 100	20.0%	15.8%	*
ILS 163 101-222 100	20.0%	14.0%	*
Over ILS 222 100	20.0%	8.8%	*
Don't know		6.2%	
Prefer not to answer 14.1%			
Region			
North	28.5%	20.4%	*
Center	40.9%	52.9%	*
Jerusalem district	12.2%	11.4%	
South	18.3%	15.3%	

CANADA	Target	Sample	+/-20%
Gender			
Male	50.0%	48.9%	
Female	50.0%	51.1%	
Age			
18-24	13.8%	13.8%	
25-34	19.7%	18.0%	
35-44	21.8%	22.4%	
45-54	22.5%	22.4%	
55-69	22.2%	23.2%	
Household income			
CAD 0-USD 34 400	20.0%	26.2%	*
CAD 34 401-49 000	20.0%	20.0%	
CAD 49 001 -65 200	20.0%	16.5%	
CAD 65 201-88 800	20.0%	13.1%	*
Over USD 88 800	20.0%	14.2%	*
Don't know		1.8%	
Prefer not to answer		8.2%	
Region			
British Columbia	13.3%	13.6%	
Other West	17.5%	15.4%	
Ontario	38.8%	41.7%	
Quebec	23.4%	22.8%	
Atlantic	7.1%	6.5%	

FRANCE	Target	Sample	+/-20%
Gender			
Male	49.4%	48.9%	
Female	50.6%	51.1%	
Age			
18-24	18.3%	16.4%	
25-34	18.1%	15.9%	
35-44	20.3%	21.1%	
45-54	19.7%	21.5%	
55-69	23.7%	25.1%	
Household income			
EUR 0-24 700	20.0%	18.7%	
EUR 24 701-32 800	20.0%	17.2%	
EUR 32 801-41 600	20.0%	17.2%	
EUR 41 601-55 500	20.0%	18.6%	
Over EUR 55 500	20.0%	11.7%	*
Don't know		4.7%	
Prefer not to answer		12.0%	
Region			
Ile de France	19.0%	20.9%	
Nord-Ouest	23.0%	24.0%	
Nord-Est	23.0%	20.2%	
Sud-Ouest	11.0%	10.4%	
Sud-Est	24.0%	24.6%	

JAPAN	Target	Sample	+/-20%
Gender			
Male	50.0%	50.3%	
Female	50.0%	49.7%	
Age			
18-24	11.5%	12.3%	
25-34	20.5%	21.9%	
35-44	19.5%	19.1%	
45-54	18.4%	20.3%	
55-69	30.0%	26.4%	
Household income			
JPY 0-2 700 000	20.0%	16.9%	
JPY 2 700 001-4 010 000	20.0%	17.0%	
JPY 4 010 001-5 350 000	20.0%	17.4%	
JPY 5 350 001-7 400 000	20.0%	18.1%	
Over JPY 7 400 000	20.0%	19.7%	
Don't know		6.7%	
Prefer not to answer		4.1%	
Region			
Hokkaido + Tohoku	11.7%	14.2%	*
Kanto	33.6%	34.5%	
Chubu	16.8%	10.0%	*
Kinki (= Kansai)	17.9%	21.3%	
Chugoku + Shikoku + Kyushu	19.9%	19.9%	

SWEDEN	Target	Sample	+/-20%
Gender			
Male	50.7%	50.4%	
Female	49.3%	49.6%	
Age			
18-24	13.2%	14.5%	
25-34	18.8%	19.5%	
35-44	21.3%	20.9%	
45-54	19.2%	19.3%	
55-69	27.5%	25.6%	
Household income			
SEK 0-242 000	20.0%	22.2%	
SEK 242 001-316 000	20.0%	19.2%	
SEK 316 001-385 000	20.0%	14.9%	*
SEK 385 001-483 000	20.0%	17.4%	
Over SEK 483 000	20.0%	18.6%	
Don't know		2.7%	
Prefer not to answer		4.7%	
Region			
Stockholm & Middle Sweden	32.0%	31.0%	
North Sweden	16.0%	15.6%	
West & South Sweden	55.0%	53.1%	

CHILE	Target	Sample	+/-20%
Gender			
Male	49.0%	48.0%	
Female	51.0%	52.0%	
Age			
18-24	19.0%	22.5%	
25-34	26.0%	26.2%	
35-44	25.0%	23.1%	
45-54	18.0%	17.9%	
55-69	12.0%	10.3%	
Household income (not used)			
CLP 0-3 210 000		20.2%	
CLP 3 210 001-4 630 000		9.1%	
CLP 4 630 001-6 180 000		11.1%	
CLP 6 180.001-8 360 000		10.2%	
CLP 8 360 001-11 420 000		33.8%	
Don't know		4.2%	
Prefer not to answer		10.8%	
Region			
Tarapaca/Antofagasta/Atacama/Coquimbo/De arica y Parinacota	11.9%	9.7%	
Valparaiso/Del Libertador/Maule	23.7%	22.2%	
Bio Bio	12.1%	12.2%	
Araucanía/Los Lagos/Aisen/Magallanes/De Los Rios	15.0%	15.9%	
Metropolitana	40.2%	40.1%	

ORGANISATION FOR ECONOMIC CO-OPERATION AND DEVELOPMENT

The OECD is a unique forum where governments work together to address the economic, social and environmental challenges of globalisation. The OECD is also at the forefront of efforts to understand and to help governments respond to new developments and concerns, such as corporate governance, the information economy and the challenges of an ageing population. The Organisation provides a setting where governments can compare policy experiences, seek answers to common problems, identify good practice and work to co-ordinate domestic and international policies.

The OECD member countries are: Australia, Austria, Belgium, Canada, Chile, the Czech Republic, Denmark, Estonia, Finland, France, Germany, Greece, Hungary, Iceland, Ireland, Israel, Italy, Japan, Korea, Luxembourg, Mexico, the Netherlands, New Zealand, Norway, Poland, Portugal, the Slovak Republic, Slovenia, Spain, Sweden, Switzerland, Turkey, the United Kingdom and the United States. The European Union takes part in the work of the OECD.

OECD Publishing disseminates widely the results of the Organisation's statistics gathering and research on economic, social and environmental issues, as well as the conventions, guidelines and standards agreed by its members.

OECD PUBLISHING, 2, rue André-Pascal, 75775 PARIS CEDEX 16
(97 2014 07 1 P) ISBN 978-92-64-21464-4 – 2014-02